PowerCLI Cookbook

Over 75 step-by-step recipes to put PowerCLI into action
for efficient administration of your virtual environment

Philip Sellers

[PACKT] enterprise
PUBLISHING   professional expertise distilled

BIRMINGHAM - MUMBAI

PowerCLI Cookbook

First published: March 2015

Production reference: 1200315

Published by Packt Publishing Ltd.
Livery Place
35 Livery Street
Birmingham B3 2PB, UK.

ISBN 978-1-78439-372-4

www.packtpub.com

Credits

Author

Philip Sellers

Reviewers

Dustin Lenz

Ajeet Singh Raina

Earl Waud

Commissioning Editor

Ashwin Nair

Acquisition Editor

Sonali Vernekar

Content Development Editor

Arwa Manasawala

Technical Editor

Menza Mathew

Copy Editor

Rashmi Sawant

Project Coordinator

Danuta Jones

Proofreaders

Simran Bhogal

Joel T. Johnson

Indexer

Rekha Nair

Graphics

Abhinash Sahu

Production Coordinator

Melwyn D'sa

Cover Work

Melwyn D'sa

About the Author

Philip Sellers is an IT enthusiast residing in Myrtle Beach, South Carolina. Philip has more than 16 years of industry experience in consulting and systems administration. Currently, he is a senior-level systems administrator with Horry Telephone Cooperative, America's largest telecommunications cooperative. Philip focuses on Microsoft and VMware software solutions along with server and SAN infrastructure. He spends a lot of time wrangling unwieldy systems and tries to tame as much as he can with automation using tools such as PowerCLI.

Philip has a bachelor of science degree in interdisciplinary studies with a minor in computer science from Coastal Carolina University and holds certifications as a VMware VCAP5-DCA and VCP5-DCV and is a Microsoft Certified IT Professional.

He maintains a technology blog at http://www.techazine.com that provides explanations and reviews of enterprise IT solutions, and he is a leader with the Myrtle Beach VMware Users Group. You may also follow him on Twitter @pbsellers.

Philip is married to his college sweetheart, Jennifer, and has two kids who keep him busy when he's not working.

Acknowledgments

I would like to thank my wonderful wife, Jennifer, for her support and patience during the project and for all of her enthusiasm and encouragement. I enjoy every day of our adventure together. I would also like to thank my children, Peyton and Jake. No dream is too big for you to dream and if you put your mind to it, step by step (or chapter by chapter) you can achieve anything you set your mind to. You three are the reasons why I do what I do and I love you all.

To my mom and dad, thank you for always supporting me, listening, and encouraging me, no matter how crazy the idea was. Thank you dad for teaching me to troubleshoot; you prepared me in ways you'll never know because you taught me to solve problems.

A special thanks to my technical reviewers; I appreciate the valuable work you did. All of your notes and suggestions helped make this book the best possible resource it could be. Thank you to the Packt Publishing team who worked on this book. Thank you to my managers and executive staff at Horry Telephone Cooperative for allowing me to pursue this opportunity.

About the Reviewers

Dustin Lenz is currently an MTS IT systems engineer with a large semiconductor manufacturer. Dustin earned a bachelor of science degree in computer technology from Ball State University, Muncie, Indiana, and has earned his certification, VCP, from VMware for datacenter technologies.

Ajeet Singh Raina is a senior systems engineer at Dell R&D. He has received a certification in VMware Certified Professional (VCP 4.1) and has more than 7 years of experience working on open source and virtualization platforms. He was a part of the VMQA GOS Validation Team at VMware India and validated all flavors of operating systems on ESXi 4.1 and 5.0. He is currently working with Enterprise Solutions Group at Dell and has a solid understanding of a diverse range of IT infrastructure, systems management, systems integration, and quality assurance.

Ajeet has a great passion for open source technologies (Linux, Hadoop, and OpenStack). He likes providing tech-talks and technical consultations on the latest open source software and has a habit of sharing it through blogs and wikis. He can be reached at http://collabnix.com

This book would not have been a success without the direct and indirect help from many people. Thanks to my wife and my 5-year-old kid for putting up with me for all the missing family time and for providing me with love and encouragement throughout the writing period. Thanks to my parents and family members for all the love, guidance, and encouragement during the tough times.

Thanks to all my past and present colleagues and mentors at VMware and Dell for the insightful discussions and the knowledge they shared with me.

Earl Waud is a virtualization development professional with more than 7 years of focused industry experience in creating innovative solutions for hypervisor provisioning, management, and automation. He is an expert in aligning engineering strategy with organizational vision and goals, and delivering highly scalable and user friendly virtualization environments.

With more than 18 years of experience in developing customer facing and corporate IT software solutions, he has a proven track record of delivering high-caliber and on-time technology solutions that have a significant impact on business results.

Earl currently lives in San Diego, California. He is blessed with a beautiful wife, Patti, and three amazing daughters, Madison, Daniella, and Alexis.

Currently, Earl is a senior systems engineer with Intuit Inc., a company that creates business and financial management solutions that simplify the business of life for small businesses, consumers, and accounting professionals.

Earl can be reached online at `http://sandiegoearl.com`.

I would like to thank my wonderful family for allowing me to spend some of my precious family time to review this book. I love and appreciate you all, and I know I am truly blessed to be part of this family.

---

www.PacktPub.com

Support files, eBooks, discount offers, and more

For support files and downloads related to your book, please visit http://www.PacktPub.com.

Did you know that Packt offers eBook versions of every book published, with PDF and ePub files available? You can upgrade to the eBook version at http://www.PacktPub.com and as a print book customer, you are entitled to a discount on the eBook copy. Get in touch with us at service@packtpub.com for more details.

At http://www.PacktPub.com, you can also read a collection of free technical articles, sign up for a range of free newsletters and receive exclusive discounts and offers on Packt books and eBooks.

https://www2.packtpub.com/books/subscription/packtlib

Do you need instant solutions to your IT questions? PacktLib is Packt's online digital book library. Here, you can search, access, and read Packt's entire library of books.

Why subscribe?

- ▶ Fully searchable across every book published by Packt
- ▶ Copy and paste, print, and bookmark content
- ▶ On demand and accessible via a web browser

Free access for Packt account holders

If you have an account with Packt at http://www.PacktPub.com, you can use this to access PacktLib today and view nine entirely free books. Simply use your login credentials for immediate access.

Instant updates on new Packt books

Get notified! Find out when new books are published by following @PacktEnterprise on Twitter or the *Packt Enterprise* Facebook page.

Table of Contents

Preface

VMware PowerCLI offers a compelling command-line alternative to the point-and-click administration of vSphere and vCloud Director. As virtualization has become mainstream and deployments begin to sprawl, the simple commands of PowerCLI allows faster administration by executing tasks on groups of objects in the virtual environment.

Since PowerCLI follows a very logical pattern, it can be quickly adopted, making it the first choice for many vSphere administrators. However, with simplicity, it also combines extensibility to allow users to build their own functions and modules to solve specific problems not addressed by out-of-box functionalities.

What this book covers

Chapter 1, Configuring the Basic Settings of an ESXi Host with PowerCLI, covers the configuration of a fresh installation of VMware ESXi on a host system.

Chapter 2, Configuring vCenter and Computing Clusters, teaches you how to perform a basic vCenter configuration and add multiple ESXi hosts into a cluster with vSphere features, such as Dynamic Resource Scheduler (DRS) and High Availability (HA).

Chapter 3, Managing Virtual Machines, provides you with many of the common tasks needed to manage virtual machines from PowerCLI, including deploying and cloning virtual machines, changing hardware settings on virtual machines, and reloading inaccessible virtual machines in vCenter.

Chapter 4, Working with Datastores and Datastore Clusters, introduces the PowerCLI cmdlets needed to create and manage datastores and datastore clusters for individual ESXi hosts or vSphere clusters.

Chapter 5, Creating and Managing Snapshots, covers cmdlets and routines to work with snapshots on virtual machines, how to manage and report on snapshots before they become problems, and uses the topic to teach you how to write your own function in PowerCLI that can be reused easily. This chapter also covers how to take your code and schedule it to run with defined triggers using native PowerShell commands.

Chapter 6, Managing Resource Pools, Reservations, and Limits for Virtual Machines, covers the topic of creating and managing resource pools and their associated settings that include reservations and limits both at a pool and virtual machine level.

Chapter 7, Creating Custom Reports and Notifications for vSphere, teaches you how to use many of the native PowerShell features for reporting and leveraging those with PowerCLI cmdlets to create custom reports and notifications.

Chapter 8, Performing ESXCLI and in-guest Commands from PowerCLI, works with the advanced topics of using ESXCLI, an alternative command-line administration tool, from within PowerCLI to access and manage settings that are not natively accessible from PowerCLI. This chapter also covers some of the basics of performing in-guest commands invoked from PowerCLI.

Chapter 9, Managing DRS and Affinity Groups PowerCLI, is built on everything covered in the previous chapters to discuss managing the vSphere DRS features from PowerCLI by building your own functions and modules to alter the group memberships of DRS groups and keep the membership updated per defined rules.

Chapter 10, Working with vCloud Director from PowerCLI, changes gears and covers managing vCloud Director and vCloud deployments in multi-tenanted environments.

Appendix, Setting up and Configuring vCloud Director, covers certain installation tips and techniques.

What you need for this book

To create and perform the commands created in the recipes of this cookbook, you will need:

- VMware vSphere PowerCLI
- Windows PowerShell 2.0 or 3.0
- VMware vCenter Server
- VMware ESXi hosts (physical or nested virtual)
- VMware vCloud Director and vShield Manager

This book was written and tested against PowerCLI versions 5.5, 5.8, and 6.0, and utilizes PowerShell 3.0.

Windows PowerShell 2.0 or 3.0 are distributed as part of the Windows Management Framework and are available for free from `http://www.microsoft.com`. VMware vSphere PowerCLI and the VMware Hypervisor (ESXi) are available for free from `http://www.vmware.com`. You can obtain a 60-day trial license for vSphere that cover ESXi and vCenter Server in order to enable advanced features and management. vCloud Director is available as a trial with a streamlined virtual appliance for evaluation purposes from `http://www.vmware.com`.

Who this book is for

This book is written for readers with a basic, working knowledge of PowerCLI, a command-line tool for managing vSphere and vCloud environments that is based on PowerShell. The book is written in a recipe format, which means that each chapter approaches a topic of vSphere or vCloud administration and walks you through step-by-step commands to handle the common tasks. Each recipe is built on the previous recipes that allow you to learn how to take basic commands and combine them into functions and modules in order to automate tasks for your environment, making your job easier.

It is assumed that you have a working understanding of VMware vSphere, both ESXi and vCenter Server, and the experience with vCloud Director might help you with the chapter focused on this topic. The book is written so that you can go beyond simple commands in PowerCLI and unleash the potential of more complex series of commands that handle real work problems. It is impossible to cover every possible use for PowerCLI, but the book covers some topics in representative ways and gives you techniques to apply to any other need you might encounter.

Conventions

In this book, you will find a number of styles of text that distinguish between different kinds of information. Here are some examples of these styles, and an explanation of their meaning.

Code words in text, database table names, folder names, filenames, file extensions, pathnames, dummy URLs, user input, and Twitter handles are shown as follows: "To check the version you are running, open a PowerCLI prompt and run `Get-PowerCLIVersion`."

Any command-line input or output is written as follows:

```
Set-PowerCLIConfiguration -InvalidCertificateAction Ignore -Scope
  Session -Confirm:$false
```

New terms and **important words** are shown in bold. Words that you see on the screen, in menus or dialog boxes for example, appear in the text like this: "Open the **Organizations** section under **Manage & Monitor** and select an organization."

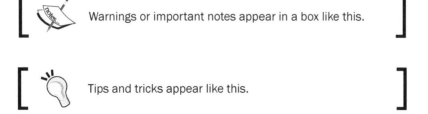

Warnings or important notes appear in a box like this.

Tips and tricks appear like this.

Reader feedback

Feedback from our readers is always welcome. Let us know what you think about this book—what you liked or may have disliked. Reader feedback is important for us to develop titles that you really get the most out of.

To send us general feedback, simply send an e-mail to feedback@packtpub.com, and mention the book title via the subject of your message.

If there is a topic that you have expertise in and you are interested in either writing or contributing to a book, see our author guide on www.packtpub.com/authors.

Customer support

Now that you are the proud owner of a Packt book, we have a number of things to help you to get the most from your purchase.

Downloading the example code

You can download the example code files for all Packt books you have purchased from your account at http://www.packtpub.com. If you purchased this book elsewhere, you can visit http://www.packtpub.com/support and register to have the files e-mailed directly to you.

Errata

Although we have taken every care to ensure the accuracy of our content, mistakes do happen. If you find a mistake in one of our books—maybe a mistake in the text or the code—we would be grateful if you would report this to us. By doing so, you can save other readers from frustration and help us improve subsequent versions of this book. If you find any errata, please report them by visiting http://www.packtpub.com/submit-errata, selecting your book, clicking on the **Errata Submission Form** link, and entering the details of your errata. Once your errata are verified, your submission will be accepted and the errata will be uploaded on our website, or added to any list of existing errata, under the Errata section of that title. Any existing errata can be viewed by selecting your title from http://www.packtpub.com/support.

Piracy

Piracy of copyright material on the Internet is an ongoing problem across all media. At Packt, we take the protection of our copyright and licenses very seriously. If you come across any illegal copies of our works, in any form, on the Internet, please provide us with the location address or website name immediately so that we can pursue a remedy.

Please contact us at copyright@packtpub.com with a link to the suspected pirated material.

We appreciate your help in protecting our authors, and our ability to bring you valuable content.

Questions

You can contact us at questions@packtpub.com if you are having a problem with any aspect of the book, and we will do our best to address it.

1
Configuring the Basic Settings of an ESXi Host with PowerCLI

In this chapter, you will cover the following recipes:

- ▶ Connecting to an ESXi host or a vCenter instance
- ▶ Getting the VMware host object
- ▶ Joining an ESXi host into Active Directory
- ▶ Enabling services and setting security profiles
- ▶ Setting network configuration
- ▶ Creating datastores on an ESXi host
- ▶ Configuring syslog settings on a host
- ▶ Joining an ESXi host to vCenter
- ▶ Creating a configuration script to set all properties uniformly

Introduction

Initially, automation doesn't save time. To get the benefits of automation, you must invest the time upfront to create scripts that you'll use time and again. In this chapter, you'll take your first ESXi host that has been installed with an IP address configured on it, and you will configure the server continually from there. This chapter will take an administrator through the basic configuration tasks needed to perform the initial configuration, join them to vCenter, and get it into an operational state. At the end of this chapter, all of these steps will build into a scripted configuration that can be executed against new hosts in the future.

Connecting to an ESXi host or a vCenter instance

To begin working with PowerCLI, you must first have PowerShell installed and available on the system on which you will run PowerCLI. PowerShell is a part of the Windows Management Framework and it ships with Windows client and server versions. PowerCLI extends PowerShell with commands to administer VMware environments. With PowerShell installed, you will need to obtain PowerCLI from `http://www.vmware.com`. The specific link is listed in the *See also* section of this recipe.

Once you have installed PowerCLI, you will need to build an ESXi host for this recipe. All that is required is a fresh ESXi installation from the ISO or DVD image distributed by VMware. Once installed, set an IP address on an accessible network using the console screens of the new ESXi host. The network address should be accessible from your PowerCLI workstation.

With the assumption that your ESXi host is built, the first step to administer VMware environments in PowerCLI is to connect to the ESXi host or to a vCenter server. In this chapter, you will focus on how to configure a single ESXi host. In the next chapter, you will focus on how to configure a vCenter Server and a vSphere cluster of ESXi hosts.

Getting ready

To begin, you only need to launch PowerCLI from its shortcut on the desktop or from the Start Menu. If you already had PowerCLI previously installed, you will want to check the version number to ensure that the cmdlet references throughout the book are available to you. Each version of PowerCLI builds additional native cmdlets and functionalities. To check the version you are running, open a PowerCLI prompt and run `Get-PowerCLIVersion`.

The recipes in this book are built and tested using VMware PowerCLI 5.5 Release 1 and have also been tested with VMware PowerCLI 5.8 Release 1, which accompanies vCloud Suite 5.8. The recipes also work on vSphere and PowerCLI 6.0 with any differences noted.

How to do it...

1. At the PowerCLI prompt, you will execute the `Connect-VIServer` cmdlet as follows:

    ```
    Connect-viserver <hostname or IP>
    ```

2. When executed, the code will attempt to perform a single sign-on into the ESXi host, but unless your username is `root` and you set the same password locally and on ESX, single sign-in will fail. You will be prompted with a normal Windows login window, which is displayed as follows, and you should log in with the `root` username and the password you specified during your ESXi installation:

3. Once you successfully log in to the ESXi host, a confirmation message will be displayed with the name or IP address of the ESXi host you connected to, the port, and the user you've connected to, as shown in the following example:

```
Name                          Port  User

----                          ----  ----

192.168.0.241                 443   root
```

4. At this point, the PowerCLI session is connected to a host and ready to execute work.

How it works...

The `Connect-VIServer` cmdlet is the simplest kind of cmdlet in PowerCLI. This cmdlet initiates a connection to the vCenter or ESXi web services to allow additional commands to be passed to the server and be executed.

The `Connect-VIServer` cmdlet requires only the name of the host to which you want to connect. There are additional parameters that you might pass to the cmdlet, such as the protocol (HTTP or HTTPS), the username, and the password. If you prefer not to keep your password in plain text, you can also pass a `PSCredentials` object. The `PSCredentials` object contains login data to authenticate. For more information about the `PSCredentials` objects, type `get-help about_server_authentication`.

Once you execute the cmdlet, a warning will be displayed in yellow, similar to the following one:

The warning is displayed because the certificate installed on the ESXi host is self-signed and untrusted by the computer you are connecting from. Changing an SSL certificate on ESXi hosts will be covered later in the book, but the warning can be ignored this time. The cmdlet will continue to execute even though the warning is displayed.

You can also prevent the invalid certificate errors by running the following PowerCLI cmdlet that changes the action when an invalid certificate is encountered:

```
Set-PowerCLIConfiguration -InvalidCertificateAction Ignore -Scope
    Session -Confirm:$false
```

There's more...

If you choose to join the ESXi host to Active Directory, your PowerCLI session performs a single sign-in. PowerCLI uses the credentials of your current Windows session to log in against the ESXi host or vCenter server if your account has access to the server. If your account does not have access to the server that it is attempting to connect to, a login box will be presented like our example, shown in the preceding screenshot, in this recipe.

See also

- ▶ The *Joining an ESXi host into Active Directory* recipe
- ▶ The *Setting permissions on vCenter objects* recipe in *Chapter 2, Configuring vCenter and Computing Clusters*
- ▶ VMware PowerCLI Documentation Center and Installation Download https://www.vmware.com/support/developer/PowerCLI/

Getting the VMware host object

Cmdlets become available to manage a host after we connect to that host to manage it. The first concept that you will need to become aware of are PowerShell objects. Objects are defined as data obtained from commands that run in PowerShell and PowerCLI. To perform configuration on an ESXi host, the commands that you run will need a host object, which is specified.

In this recipe, you will learn how to obtain a VMHost object.

Getting ready

To begin with, open a PowerCLI window and connect to an ESXi host or a vCenter instance.

How to do it...

1. PowerCLI is straightforward. To retrieve an ESXi host object, just run the following command line:

   ```
   Get-VMHost
   ```

2. After running the Get-VMHost cmdlet, an object that contains one or more ESXi hosts is returned. You are connecting to a single ESXi host in this example and running Get-VMHost that returns the host object with a single host. If you were connecting against a vCenter instance, Get-VMHost (with no other arguments) would return an object that contains all of the hosts managed by vCenter. When running against vCenter, you can specify a filter with the Get-VMHost cmdlet in order to find one or more hosts that match the specified pattern:

   ```
   Get-VMHost esxhost*
   ```

   ```
   Get-VMHost VMHOST1
   ```

3. Instead of calling the Get-VMHost cmdlet each time, you need to get the ESXi host. You can store the host object in a variable. PowerShell variables are specified using $ followed by a name. The following is an example of our ESXi host:

   ```
   $esxihost = Get-VMHost
   ```

How it works...

To learn more about the VMHost object, you can use the Get-Member cmdlet with the variable you have just defined. To use Get-Member, you will call the VMHost object by typing the $esxihost variable. Then, you pipe the object into the Get-Member cmdlet as follows:

```
$esxihost | Get-Member
```

PowerCLI is an extension of PowerShell that is used specifically for VMware product management. PowerShell is an object-based language that uses the concept of encapsulating both data and operations within an object data type, which is a familiar object-oriented programming concept. Objects have defined data areas and can include functions that perform operations on the data in the object.

The output from the cmdlet shows all of the data contained in the `Property` elements in the object. The object also includes a number of methods. These methods are used to manipulate the data in the object. The output of the preceding command is shown in the following screenshot:

You can call a method by using a dot notation (.) and by calling the method name followed by parenthesis, such as in the following example:

```
$esxihost.ConnectionState.ToString()
```

In the preceding example, the `State` property is an object inside the `VMHost` object, but the `ToString()` method converts the output to a string.

Now that the ESXi host object is stored in a variable, you can proceed with other cmdlets for configuration and run them using the host object to perform the configuration.

There's more...

`Get-VMHost` has other applications other than just returning the `VMHost` object to use. Like all other `Get-` cmdlets, this cmdlet can be used to find a host in a particular configuration or state. You can use `Get-VMHost` to find hosts assigned to a particular location in vCenter using the `-Location` parameter. You might want to find hosts that have been assigned a particular tag in vSphere using the `-Tag` parameter or you might want to find the host running a particular VM with the `-VM` parameter. Another interesting use case is specifying the `-Datastore` parameter to find all of the hosts that have a particular datastore connected.

`Get-VMHost` is just one of the many cmdlets that work with `VMHost` objects. Others will be explored in *Chapter 2, Configuring vCenter and Computing Clusters*.

See also

- The *Setting up folders to organize objects in vCenter* recipe in *Chapter 2, Configuring vCenter and Computing Clusters*
- The *Creating basic reports of VM properties using VMware Tools and PowerCLI* recipe in *Chapter 3, Managing Virtual Machines*

Joining an ESXi host into Active Directory

As mentioned in the connecting section, joining an ESXi host to Active Directory offers the ability to connect to the host without entering the credentials for administrators. Active Directory is a Windows implementation of **Lightweight Directory Access Protocol** (**LDAP**). It contains accounts for users, computers, and groups. It runs on a Windows Server that has the Active Directory role installed and that has been "promoted" to become a domain controller. To perform this recipe, you will need at least one Active Directory server available on the network with the ESXi host.

Seamless authentication is one of the biggest reasons to join a host to Active Directory. However, beyond single sign-on, once the ESXi host is connected to Active Directory, groups in the directory can be leveraged to grant permissions to the ESXi host. If you do not have Active Directory installed and do not wish to, you can skip this recipe and move on to other topics of host configuration without any impact to future recipes.

Getting ready

PowerCLI has `Get-VMHostAuthentication` and `Set-VMHostAuthentication`, two cmdlets to deal with host authentication. To get ready to set up authentication, open a PowerCLI window and connect to a single ESXi host.

How to do it...

1. Because the cmdlets require a `VMHost` object, you'll again be using `Get-VMHost` to either populate a variable or to pipe an object to the next object. The first step is to obtain a `VMHost` object for our target ESXi host. This can be done using the following command line:

    ```
    $esxihost = Get-VMHost 192.168.0.241
    ```

2. Once you have your `VMHost` object, you can look at setting up the authentication. The `Set-VMHostAuthentication` cmdlet needs to be executed. The cmdlet requires several parameters to join an ESXi host to the domain. The syntax needed is displayed as follows:

    ```
    $esxihost | Get-VMHostAuthentication | Set-VMHostAuthentication
    -JoinDomain -Domain domain.local -user username -password *****
    ```

3. Executing the cmdlet will prompt you to confirm that you want to join this host to the domain specified. If your answer is `Y`, the cmdlet will continue and execute the operation as follows:

    ```
    Perform operation?

    Joining VMHost '192.168.0.241' to Windows Domain 'domain.local'.

    [Y] Yes   [A] Yes to All   [N] No   [L] No to All   [S] Suspend   [?]
    Help

    (default is "Y"):Y


    Domain              DomainMembershipStatus      TrustedDomains

    ------              ----------------------      ---------------

    DOMAIN.LOCAL        Ok
    ```

How it works...

One of the first things you will notice about this recipe is that there is an extra `Get-VMHostAuthentication` cmdlet in the middle of the command line. Why does it need to perform `Get` before performing `Set`? It would seem that you can simply pipe the `VMHost` object into cmdlet to specify your target host and the cmdlet will execute the function. But as you try that, using the following command line, PowerCLI displays an error, as shown in the following screenshot:

```
$esxihost | Set-VMHostAuthentication -JoinDomain -Domain domain.local
-user username -password *****
```

In this case, the cmdlet looks for a VMHostAuthentication object and not a VMHost object, so an error is displayed. If you go back and simply execute the Set-VMHostAuthentication cmdlet as follows, it will prompt you for a VMHostAuthentication object and wait for an input:

```
Set-VMHostAuthentication -JoinDomain -Domain domain.local -user username
-password *****
```

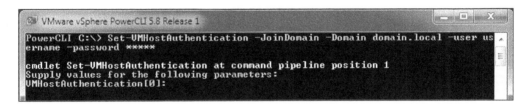

This is where the Get-VMHostAuthentication cmdlet gets added. It retrieves the VMHostAuthentication object from the host you targeted since this cmdlet accepts the VMHost object as a piped input.

The Get-Help cmdlet for Set-VMHostAuthentication also shows that the cmdlet expects a VMHostAuthentication object to be passed as a parameter for the cmdlet. By executing the cmdlet with all of its parameters and no piped input, you also learned that you can debug and learn what input the cmdlet is expecting and missing.

There's more...

The same cmdlets can also be used to remove a host from a domain, if needed. The -LeaveDomain parameter is a part of the Set-VMHostAuthentication cmdlet and allows this need.

In addition to setting up an ESXi host to accept Active Directory authentication, PowerCLI also provides a number of cmdlets to add local users, groups, and permissions inside a single ESXi host. The `New-VMHostAccount` cmdlet is used to create new users on an ESXi system. The same cmdlet previously allowed the creation of groups, but this functionality was removed with ESXi 5.1. There is a `Set-VMHostAccount` cmdlet to change accounts and group memberships, and a `Remove-VMHostAccount` cmdlet to remove a user or a group.

See also

> ▸ The *Setting permissions on vCenter objects* recipe in *Chapter 2, Configuring vCenter and Computing Clusters*

Enabling services and setting security profiles

ESXi hosts enable a few services by default, but there are some additional services that are installed but blocked. In some cases, you might want to enable SSH on the host. However, since VMware does not recommend enabling SSH and will display a warning. You can set an advanced setting to disable this warning.

Getting ready

To begin with, you should open a PowerCLI prompt and connect to an ESXi or vCenter host. You will also want to store a `VMHost` object in a variable named `$esxihost`.

How to do it...

1. The first step is to get the list of available services from a VMware host. To do this, you use the `Get-VMHostService` cmdlet and pass the `VMHost` object into the cmdlet as follows:

    ```
    $esxihost | Get-VMHostService
    ```

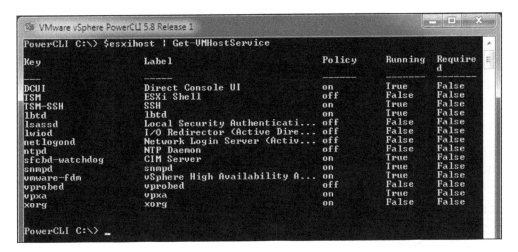

2. The output of the preceding cmdlet will display a list of the available services on the ESXi host along with its policy (whether it is set on or off by default) if it's running. The label is a friendly identifier to find the service you want to configure, but the key is the piece of data you will use to return the single service you want.

3. In this case, we're looking to configure the service with the `TSM-SSH` key. To scope the results down to that one service in the object, you will use a PowerShell `where` clause as follows:

   ```
   $esxihost | Get-VMHostService | where { $_.key -eq "TSM-SSH" }
   ```

4. Now that you have it scoped down to a single service, you pass this object into the `Set-VMHostService` cmdlet with the desired policy of `On` as follows:

   ```
   $esxihost | Get-VMHostService | where { $_.key -eq "TSM-SSH" } |
   Set-VMHostService -Policy "On"
   ```

5. At this point, you have configured the host to autostart the service on boot, but the service is still not running in the current boot. To do this, you will instead use the `Start-VMHostService` cmdlet. Again, you have to pass in the `VMHostService` object for SSH (or any other service that you choose).

```
$esxihost | Get-VMHostService | where { $_.key -eq "TSM-SSH" } |
Start-VMHostService
```

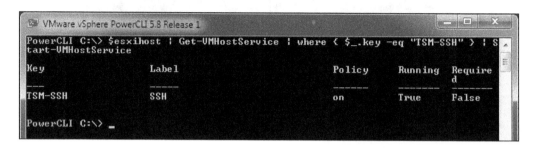

6. With the service running, vSphere displays the warning that you have enabled SSH. This will leave your host showing in a warning state as long as the service is running; however, VMware does allow you to suppress this warning, but this is set through an advanced setting. To set this, you need to execute the following cmdlet:

```
$esxihost | Get-AdvancedSetting –Name UserVars.
SuppressShellWarning | Set-AdvancedSetting –value 1
```

7. When executed, the preceding command line will prompt you to confirm the settings. This confirmation can be suppressed using the `–Confirm:$false` common parameter, which is useful in scripts:

```
$esxihost | Get-AdvancedSetting –Name UserVars.
SuppressShellWarning | Set-AdvancedSetting –value 1 –
Confirm:$false
```

How it works...

For configuring host services, the native cmdlets follow the expected pattern of Get and Set functionality in PowerCLI. `Get-VMHostService` expects a VMHost object as the input which is logical since these host services exist within the scope of a host. Once you get the host service by name and store it in a variable or pass it as an object in the pipeline, you can easily set the settings to the desired configuration. In addition to Get and Set cmdlets, you also have Start and Stop cmdlets. The Start and Stop cmdlets are more specific to this use case since we're dealing with host services and there is a specific need to start or stop them in addition to configuring them. The Start and Stop cmdlets also accept the HostService objects as inputs, just like the `Set-VMHostService` cmdlet.

In the specific use case of the SSH Server service, it causes a warning to be displayed to the client. To disable this warning from been displayed, you can use an advanced setting named `UserVars.SupressShellWarning`. While this is not recommended for production systems, there are plenty of use cases where SSH is needed and is helpful in lab environments, where you might want to configure the setting.

There's more...

The cmdlet to start the SSH service can be easily adapted beyond the illustrated use case with the use of a `ForEach` loop. For troubleshooting and configuration, you might need to enable SSH in order to tail a log file or to install a custom module. In these cases, starting SSH in bulk might be handy. To do this, you take the preceding code and wrap it in the loop. An example of a connection to a vCenter host, a variable with multiple VMHost objects returned, and a loop to step through and start SSH on each is shown as follows:

```
Connect-VIServer vcenterhost.domain.local
$esxihosts = Get-VMHost
foreach ($esxihost in $esxihosts) {
$esxihost | Get-VMHostService | where { $_.key -eq
"TSM-SSH" } | Start-VMHostService
}
```

This quickly allows you to turn on SSH for temporary use. Following a reboot, the service will no longer be running and you can easily change the preceding code to be a `Stop-VMHostService` cmdlet and turn off the service in bulk.

Setting network configuration

One of the first things to be completed against a new ESXi installation is network configuration. Network configuration consists of several things on an ESXi host—first would be to configure the additional management interfaces of the host for VMotion, Fault Tolerance logging, vSphere Replication, and VSAN traffic.

Getting ready

To begin this recipe, you will need to open a PowerCLI window, connect to an ESXi host, and load a `VMHost` object into a variable. The example uses `$esxihost` as the variable for the `VMHost` object.

On installation, ESXi has a single **Network Interface Card** (**NIC**) labeled `eth0` that is connected to a VMware Standard—**vSwitch**. The vSwitch has two port groups created: one labeled **Management Network** for management traffic and the other is labeled **VM Network**. The Management Network is a vmkernel port with the IP defined on the console attached to it.

In this example, our host contains six 10 Gigabit NICs that will connect the host to the network. You will define two additional vSwitches with two physical ports attached to each for redundancy. The additional vSwitches will handle storage and replication traffic on one and VM traffic on the other.

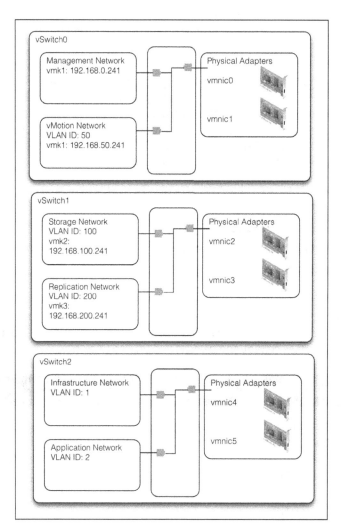

Best practices of vSphere networking are far beyond the scope of this book. The network layout shown in the preceding diagram is not an endorsement of a particular layout and is for illustration purposes to show the PowerCLI cmdlets used to configure networking on ESXi.

How to do it...

1. To begin with, let's get an idea of the network layout that is in place, by default. From a default install, there is a single virtual switch named `vSwitch0`. The first cmdlet shows you the properties of this virtual switch and the second shows you the port groups associated with that vSwitch. To do this, review the output of the two PowerCLI cmdlets:

   ```
   $esxihost | Get-VirtualSwitch

   $esxihost | Get-VirtualPortGroup -VirtualSwitch vSwitch0
   ```

2. The first thing to be completed is to remove the default VM Network port group, since it's not the best practice to have Virtual Machine traffic on the management ports, and this default port group is not a part of the design you outlined for this configuration:

   ```
   $esxihost | Get-VirtualPortGroup -Name "VM Network" | Remove-
   VirtualPortGroup -Confirm:$false
   ```

3. The preceding command combines the `Get-VirtualPortGroup` and `Remove-VirtualPortGroup` cmdlets to change the confirmation. When executed, you will receive either a confirmation or an error. If the port group is connected to or in use by a VM, you will receive an error message. Once you remove the VM Network port group, the next step is to add an additional vmkernel port that will be used for vMotion.

> While this is outside the scope of this book, there are many different ideas for the best design of VMware networking. Most administrators agree that Management traffic and vMotion traffic should be separated, but with increasing speeds and capabilities of NICs today, it's common to see them sharing the same virtual switch. Administrators will set the Management traffic to be active on the first NIC and vMotion to be active on the second NIC. The two traffic streams will only be on the same NIC in a failover situation.

4. In our design, you will set Management and vMotion to be collocated on the same switch. To do this, use the `New-VMHostNetworkAdapter` cmdlet and pass in the name of the port group, the virtual switch, and the IP information. You will also pass in a parameter to specify that this vmkernel port should be used for VMotion as follows:

   ```
   $esxihost | New-VMHostNetworkAdapter -PortGroup "VMotion
   Network" -VirtualSwitch vSwitch0 -IP 192.168.50.241 -SubnetMask
   255.255.255.0 -VMotionEnabled $true
   ```

5. In our design, although vMotion and Management traffic exist on the same vSwitch, the traffic will be separated using active and standby links on each port group. This is done by changing the **NIC Teaming Policy** with the `Set-NicTeamingPolicy` cmdlet. You can see in the following two commands that the active and standby NIC assignments are opposite between the two port groups:

```
$esxihost | Get-VirtualPortGroup -Name "Management Network" | Get-
NicTeamingPolicy | Set-NicTeamingPolicy -MakeNicActive vmnic0 -
MakeNicStandby vmnic1
```

```
$esxihost | Get-VirtualPortGroup -Name "VMotion Network" | Get-
NicTeamingPolicy | Set-NicTeamingPolicy -MakeNicActive vmnic1 -
MakeNicStandby vmnic0
```

6. The port group is automatically created and the vmkernel/host port is created for our vMotion network, but it's on the wrong VLAN. Our vMotion traffic is on a different VLAN, so you need to set this on the port group as follows:

```
$esxihost | Get-VirtualPortGroup -Name "VMotion Network" | Set-
VirtualPortGroup -VlanID 50
```

7. The next step is to create a new virtual switch with its own uplinks on vmnic2 and vmnic3, as shown in our diagram. To confirm that the physical NICs exist, you can run the following cmdlet:

```
$esxihost | Get-VMHostNetworkAdapter
```

The `Get-VMHostNetworkAdapter` cmdlet displays all of the vmkernel ports along with all of the physical NICs present on the host.

8. After confirming the NIC, you will run the `New-VirtualSwitch` cmdlet to provision the new virtual switch. This cmdlet provisions the vSwitch with its uplinks, but it's currently an island with no connectivity for Management or virtual servers:

```
$esxihost | New-VirtualSwitch -Name vSwitch1 -Nic vmnic2,vmnic3
```

9. The next step is to create vmkernel ports for storage traffic and replication traffic. These are created in the same way as the VMotion Network we provisioned earlier:

```
$esxihost | New-VMHostNetworkAdapter -PortGroup "Storage
Network" -VirtualSwitch vSwitch1 -IP 192.168.100.241 -SubnetMask
255.255.255.0 -VsanTrafficEnabled $true
```

```
$esxihost | Get-VirtualPortGroup -Name "Storage Network" | Set-
VirtualPortGroup -VlanID 100
```

```
$esxihost | New-VMHostNetworkAdapter -PortGroup "FT Logging
Network" -VirtualSwitch vSwitch1 -IP 192.168.200.241 -SubnetMask
255.255.255.0 -FaultToleranceLoggingEnabled $true
```

```
$esxihost | Get-VirtualPortGroup -Name "FT Logging Network" | Set-
VirtualPortGroup -VlanID 200
```

10. Again, you want to make sure that our Storage Traffic and Fault Tolerance traffic don't end up competing for bandwidth. Therefore, you will assign one port group to be active on one uplink and the other port group to be active on the second uplink. This is done again with the `Set-NicTeamingPolicy` cmdlet:

```
$esxihost | Get-VirtualPortGroup -Name "Storage Network" | Get-
NicTeamingPolicy | Set-NicTeamingPolicy —MakeNicActive vmnic2 -
MakeNicStandby vmnic3

$esxihost | Get-VirtualPortGroup -Name "FT Logging Network" | Get-
NicTeamingPolicy | Set-NicTeamingPolicy —MakeNicActive vmnic3 -
MakeNicStandby vmnic2
```

11. The final step of our network provisioning is to create new port groups for Virtual Machine traffic. You have set all of the virtual machine traffic to its own vSwitch and uplinks in the design outlined. The first step is to create the virtual switch like you did for vSwitch1 as follows:

```
$esxihost | New-VirtualSwitch -Name vSwitch2 -Nic vmnic4,vmnic5
```

12. Once the virtual switch is created, you can create two port groups on the virtual switch. However, in this case, `New-VirtualPortGroup` doesn't allow any pipeline input, so you will need to specify the server as a parameter instead of passing it through the pipeline:

```
New-VirtualPortGroup -Name "Infrastructure Network" -VirtualSwitch
vSwitch2 -VLanId 1 -Server 192.168.0.241

New-VirtualPortGroup -Name "Application Network" -VirtualSwitch
vSwitch2 -VLanId 2 -Server 192.168.0.241
```

How it works...

In this example, you will work with the VMHost object to enumerate and identify the existing configuration that is put in place during the installation. From there, you remove the default VM networking configuration, you provision new virtual switches and vmkernel ports to segment traffic, and you enable certain management functions across the vmkernel ports.

While most of the configuration covered in this section deals with the initial configuration of a host, some of the concepts are repeated more often. For instance, if you have a multi-node cluster and you're adding a new virtual machine network, you'll use the New-VirtualPortGroup cmdlet often. As you have seen in previous examples, you can easily create an array of ESXi hosts—either by using Get-VMHost in vCenter or by manually specifying a list of hosts—and then connect and provision the same port group on many hosts, quickly. This would mean big time savings and less potential for manual error when compared to manually clicking on each through the GUIs to configure the new port group on each host in the cluster.

By also using the `Set-NicTeamingPolicy` cmdlet, you can set a preferred uplink port for each port group and put the other NIC into standby mode. This allows us to keep the Management and vMotion and the Storage and Fault Tolerance traffic separated so they will not cause the performance of one another to be degraded.

There's more...

In this recipe, you focused on VMware Standard vSwitches. Users with Enterprise Plus licensing also have the option of using VMware Distributed vSwitches which have their own set of cmdlets to manage and configure these advanced virtual switches.

See also

▶ The **Network Management with vSphere Distributed Switches** under **VMware vSphere 5.5 Documentation Center** page at `http://pubs.vmware.com/ vsphere-55/topic/com.vmware.powercli.ug.doc/GUID-D2C0E491-A0CB- 4799-A80D-19EA9114B682.html` or else just type `http://bit.ly/1wJs1JP`.

Creating datastores on an ESXi host

With networking, VMware has done a lot of work to ease administration with the VMware Distributed Virtual Switch. In vSphere 5.5, VMware introduced Datastore Clusters that alleviate some of the management of datastores. However, from a provisioning standpoint, the initial setup of storage is still manual and can take a lot of manual steps. Scripting this makes a lot of sense in large environments.

Datastore and storage under vSphere is also different since some operations must be performed on the raw storage device and these steps are not repeated on every host. There are three types of storage connectivity that you might need to provision: NFS, iSCSI, and Fibre Channel. For this example, you will focus on iSCSI and NFS, and you will work on provisioning storage from both. Along the way, Fibre Channel will also be discussed since its concepts overlap with iSCSI from a vSphere perspective.

Getting ready

For this example, you will need to open a PowerCLI window and connect to an ESXi host. You will also want to make sure that you have the VMHost object stored in a variable named `$esxihost`, covered in the *Getting the VMware host object* section.

How to do it...

1. The simplest of all datastores to provision is an NFS datastore. A single PowerCLI cmdlet will provision an NFS datastore. The `New-Datastore` cmdlet will take all of the input needed to provision the new datastore and make it available for use. Since NFS does not use the VMFS filesystem, there are no filesystem properties that need to be passed. To connect NFS, you just need to provide a name for vSphere to identify the datastore, a path (the export), and the host that is providing the NFS, as follows:

   ```
   $esxihost | New-Datastore -Nfs -Name DataStoreName -Path /data1/
   export -NfsHost nfsserver.domain.local
   ```

2. With this, you've got your first datastore presented and ready to host virtual machines. For NFS, this is all that is required.

3. iSCSI and Fibre Channel storage is a bit more complex to provision from a PowerCLI and vSphere perspective. Provisioning storage on either of these protocols will require additional decisions to be made when creating the datastore. iSCSI also requires additional configuration steps that are not needed with Fibre Channel. We will focus on iSCSI in this example and I will make a note of where the concepts overlap with Fibre Channel.

4. iSCSI is an IP-based storage protocol, and as such, you will need to do a bit of network configuration to set up iSCSI to work in our environment. The first thing that needs to be done is to enable iSCSI and to create a software iSCSI target, as follows:

   ```
   $esxihost | Get-VMHostStorage
   ```

5. By default, there isn't a software iSCSI target that is created. To create this, you need to expand on the previous cmdlet and set this value to `true`, as follows:

   ```
   $esxihost | Get-VMHostStorage | Set-VMHostStorage
   -SoftwareIScsiEnabled $true
   ```

6. The next step is to set the iSCSI targets using the `New-IscsiHbaTargets` cmdlet. This cmdlet requires that you pass in the iSCSI HBA as an object, so first, you retrieve the iSCSI HBA using `Get-VMHostHba` and store it in a variable and then use it with `New-IscsiHbaTargets`:

   ```
   $iSCSIhba = $esxihost | Get-VMHostHba -Type iScsi

   New-IScsiHbaTarget -IScsiHba $iSCSIhba -Address $target -ChapType
   Required -ChapName vSphere -ChapPassword Password1
   ```

> In the example, there are additional parameters for authentication. iSCSI uses **Challenge-Handshake Authentication Protocol** (**CHAP**) to authenticate sessions to the target storage. Authentication is not required and if the storage system is not configured for authentication, these parameters can be omitted. However, it's a bad practice to deploy a production storage array without authentication.

7. The final step of the initial iSCSI configuration is to bind the iSCSI HBA to a specific port. Since you created a Storage Network management port, this is the port that you want to use. To make this change and to remove any other ports, you have to use the ESXCLI interface within PowerCLI. There isn't a native PowerCLI cmdlet for this function:

```
$esxcli = Get-ESXCLI -VMHost $esxihost
$esxcli.iscsi.networkportal.add($iscsihba, $true,"vmk2")
```

8. In our case, the vmkernel port assigned to the Storage Network port group is vmk2. Using the ESXCLI interface, you can assign it to the iSCSI HBA. To confirm the change, you can use the `list()` method, as follows:

```
$esxcli.iscsi.networkportal.list()
```

9. As you will see, there are other vmkernel ports listed. In my case, vmk0; you can remove them with a simple `remove()` method, as follows:

```
$esxcli.iscsi.networkportal.remove($iscsihba,$true,"vmk0")
```

Now that the system has its targets configured, if the iSCSI array has provisioned storage to the host, it should be visible. This is the point where iSCSI and Fibre Channel converge. Since iSCSI uses the host bus adapter model that Fibre Channel invented, they work in the same way after initial configuration. You must run the NFS mount on each server and you must set up iSCSI initial configuration on each host. Scanning and formatting VMFS datastores only needs to be done from a single host for iSCSI and Fibre Channel disks since they are shared resources. This means that when scripting the steps on each host, the next few steps only need to be done on a single host in the cluster and then every host needs to be a rescan:

```
$esxihost | Get-VMHostStorage -RescanAllHBA -RescanVmfs
```

Starting with a rescan is a good idea so that your system recognizes all of the storage changes and sees all disks that have been presented. Whether you're using software or hardware iSCSI, Fibre Channel, or converged network adapters, this is the point where your hosts see its SAN disks.

10. At this point, your ESXi system doesn't have iSCSI or Fibre Channel datastores that it can use. Even though the disk is visible, it is unformatted and not ready to host VMs. To discover your disks and to enumerate the data you need to configure it, you will need to use the `Get-ScsiLun` cmdlet:

```
$esxihost | Get-ScsiLun
```

11. This returns a list of disks available to the SCSI subsystem under ESXi. The list might contain a lot of objects. You can use various properties returned by the `ScsiLun` object to identify and leverage the list for provisioning. For instance, you can scope the list using the Vendor property or by the model. For the purpose of this example, we will assume that you have a disk identified by the `iSCSIDisk` model and use that for scoping. To create a new datastore on the disk, you need the canonical name, which is also a property in the ScsiLun object:

```
$LUN = $esxihost | Get-ScsiLun | Where {$_.Model -like
"iSCSIDisk"}
```

12. In situations where you have many disks presented to a host, identification by model might not be the best. Another method would be to use the `RuntimeName` property that enumerates the HBA, controller, target, and the LUN number. For instance, if you know the LUN number you want to prepare is LUN 8 that is represented in the `RuntimeName` as `L8`, the PowerCLI to scope and return this would be as follows:

```
$LUN = $esxihost | Get-ScsiLun | Where {$_.RuntimeName -like
"*L8"}
```

13. By storing the LUN in a variable, we can verify the returned value to ensure that you have the correct object and number of objects expected before passing it into the `New-Datastore` cmdlet:

```
$esxihost | New-Datastore -Name iSCSIDatastore1 -Path $LUN.
CanonicalName -VMFS
```

14. This provisions the disk as a VMFS filestore and allows it to be used for VM storage. At this point, you can initiate a rescan on all of the ESXi hosts in the cluster and they will all see the same shared storage.

How it works...

Provisioning datastores in vSphere works differently for each type of SAN storage. NFS is simpler than iSCSI or Fibre Channel and just requires that you connect (or mount) the datastore for use on the host. Software-based iSCSI requires that you do some additional configuration on the host so that it can connect to the target array, but then iSCSI and Fibre Channel both will work in the same way with backend storage LUNs being presented to the host for consumption.

See also

▸ The *Creating and managing datastore clusters* and the *Performing Storage vMotion* recipes in *Chapter 4, Working with Datastores and Datastore Clusters*

Configuring syslog settings on an ESXi host

Booting your ESXi from SD or USB flash storage is a common scenario. However, when booting from SD and USB, ESXi does not use that storage for logging. Instead, it keeps the logs in memory, which is nonpersistent. Now that you have established a shared, persistent storage, you can point the ESXi hosts syslog functions to store the logs onto the shared disk so that it can survive a reboot or help you to troubleshoot. Even hosts booting from a local spinning disk might want to redirect their syslog onto a shared SAN drive so that it's accessible from another hosts if one of the hosts fails.

Another common use in enterprises is a centralized syslog server or a third-party log collection and analytics service, such as Splunk. Third-party services offer filters, alarms, search, and other advanced features to add context and value to the logs collected from systems.

This section will cover setting this configuration on an ESXi host.

Getting ready

To work in this section, you will need to open a PowerCLI window, connect to an ESXi host, and populate the `$esxihost` variable with a `VMHost` object.

How to do it...

1. PowerCLI provides the `Get-AdvancedConfig` cmdlet that lets us peer into the advanced settings of the ESXi host. Even in the GUI, the syslog settings for an ESXi host are set within the Advanced Configuration setting. If you enumerate all of the advanced settings and then scope for items with `syslog.global`, you will see the settings you want to adjust to set centralize logging:

    ```
    $esxihost | Get-AdvancedSetting | Where {$_.Name -like "syslog.
    global*"}
    ```

    The two settings you want to adjust are: `logDirUnique` that sets a subdirectory for each host in the cluster, and `logDir` that sets the centralized location.

2. The `logDirUnique` setting is an easy one. First, you will need to scope down to retrieve just that setting and then pipe it into the `Set-AdvancedSetting` cmdlet:

```
$esxihost | Get-AdvancedSetting | Where {$_.Name -like "Syslog.
global.logDirUnique"} | Set-AdvancedSetting -value $true
-Confirm:$false
```

3. The second directory takes a bit more configuration. The `logDir` setting is a string that defines a storage path. So in our case, you need to figure out which datastore we're going to locate the syslog files onto. The VMFS datastore is identified as a bracketed name, which is followed by a path name. In the earlier example, you created a datastore called `iSCSIDatastore1` and you will now use it as our syslog global directory:

```
$esxihost | Get-AdvancedSetting | Where {$_.Name -like
"Syslog.global.logDirUnique"} | Set-AdvancedSetting -value
"[iSCSIDatastore1] syslog" -Confirm:$false
```

Alternatively, if you want to direct all log files to a centralized syslog server, you can set this setting, the `Syslog.global.logHost` value.

4. To set the syslog host value, you will use the same cmdlet used to set the previous values for syslog, except that you will alter the advanced setting used in the `Where` statement. The value should be `Syslog.global.logHost` to locate the correct value to be set:

```
$esxihost | Get-AdvancedSetting | Where {$_.Name -like "Syslog.
global.logHost"} | Set-AdvancedSetting -value " tcp://
syslogserver:514 " -Confirm:$false
```

How it works...

The vSphere Advanced Settings control the syslog functions. There are properties in the advanced settings that control how often and at what frequency to roll the log files, and in this example, where to store the global syslog directory, and whether to make a unique subdirectory for this host's log files.

The `Get-AdvancedSetting` and `Set-AdvancedSetting` cmdlets expose and allow us to set these Advanced Settings from PowerCLI.

Setting the global log directory requires the administrator to choose a datastore and a subdirectory on which to create these log files. The format of the path is set by using the bracketed datastore name and then a relative path inside the datastore. This is a path definition that vSphere understands, but it is also specific to vSphere. It uses a Linux-like path definition, but it begins inside the datastore location.

There's more...

In general, it's best to leave vSphere advanced settings with their default values unless instructed to make changes by VMware support. The vSphere advanced settings can alter the behavior of ESXi significantly and should be done with caution.

Joining an ESXi host to vCenter

Joining an ESXi host to vCenter is done from vCenter. The cmdlets for adding a host to a vCenter installation all require communication with vCenter. In this section, we'll connect to vCenter and add the host into inventory. All additional configuration to vCenter from PowerCLI will be covered in the next chapter.

Getting ready

Open a new PowerCLI window. This will ensure that no variables are populated and no open connection to an ESXi is lingering.

How to do it...

In this example, you will connect to a vCenter Server instead of directly connecting to an ESXi host. Our vCenter server has the hostname: `vcentersrv.domain.local`.

1. To connect to vCenter, use the same cmdlet that you used in the *Connecting to an ESXi host or a vCenter instance* recipe:

   ```
   $vcenter = connect-viserver vcentersrv.domain.local
   ```

    The same certificate warning might be displayed and you might be prompted to log in if your computer cannot single sign-in to the vCenter instance.

2. Once connected to vCenter, you can use the `Add-VMHost` to add the host into inventory:

   ```
   Add-VMHost -Server $vcenter -Name esxsrv1.domain.local
   -Location "Primary"
   ```

    For the purpose of this section, the value of `-Location` is assumed to be a datacenter object already created in vCenter. In the next chapter, you'll see code on how to create this datacenter object.

3. When prompted, enter the administrative account credentials for the ESXi to perform the `join` operation.

4. The host is now added to vCenter Server and can be administered by the server.

How it works...

Joining an ESXi to vCenter is a simple cmdlet to configure and complete. It simply links the ESXi into vCenter so that all of the additional configuration and control will be directed from the vCenter host.

At this point, connecting to the ESXi host will display a message in the GUI clients that shows it's being managed by vCenter and that all changes should be made through vCenter. That is mostly the case from PowerCLI too, but there might be additional times when configuration needs to be made directly against a host. One example would be to change multipathing settings for storage.

See also

▶ The *Creating a virtual datacenter in vCenter, Creating a cluster and adding ESXi hosts*, and *Setting cluster advanced features, including HA, DRS, and EVC* recipes in *Chapter 2, Configuring vCenter and Computing Clusters*

Creating a configuration script to set all properties uniformly

In this section, you are going to cover all of the cmdlets that have been covered in this chapter by bringing them together into a single script. This script will allow us to take an array of ESXi hosts identified by either their hostname or IP address and to run the full scripted configuration against them.

In many ways, this PowerCLI script will function much like a Host Profile in vSphere. Host Profiles are a configuration definition that can be created from an existing, configured host and applied on hosts to establish a desired configuration state. If hosts deviate from the configuration, the profile might be reapplied to remediate any undesired changes.

Unfortunately, Host Profiles are only available to customers with Enterprise Plus licensing. However, this PowerCLI solution will work for any vSphere customer with Essentials, Essentials Plus, Standard, or Enterprise licensing.

Getting ready

For this last recipe of the chapter, you'll most likely want to open something such as PowerShell ISE. PowerShell ISE provides you with additional tools to edit larger scripts, color code the cmdlets, and ensure that there are no syntax errors. Alternatively, you might want a text editing tool such as NotePad, NoteTab Light, Sublime Text, or NotePad++.

How to do it...

1. First things first, let's begin with pseudocode/documentation of what you want to accomplish. In between each of these sections, you will insert the code you have previously developed individually and put them into a full file:

   ```
   # Script to mass configure ESXi hosts

   # Step 1 - Store credentials for ESXi hosts

   # Step 2 - Set a list of target ESXi hosts and IP settings

   # Step 3 - Write a ForEach loop to iterate through hosts

   # Step 4 - Connect to ESXi host

   # Step 5 - Obtain a VMHost object to use for configuration

   # Step 6 - Join the ESXi system to Active Directory

   # Step 7 - Enable services on the ESXi host & set firewall

   # Step 8 - Configure the network settings

   # Step 9 - Configure NFS & iSCSI settings

   # Step 10 - Join hosts to vCenter

   # Step 11 - Rescan for storage changes

   # Step 12 - Configure persistent syslog storage
   ```

2. Since you want this script to do as much without any manual intervention, you will want to try and eliminate as many prompts as possible. Since you will be connecting to and executing commands on multiple ESXi hosts, you would normally get prompted to login each time you connect to a host. To avoid this, you can store the credentials in a variable and pass them to each `connect-viserver` cmdlet.

   ```
   # Step 1 - Store credentials for ESXi hosts

   $esxiCreds = Get-Credential
   ```

When you first covered connecting to ESXi servers from PowerCLI, you experienced the login box for the host. The Get-Credentials cmdlet causes the same action but returns a credentials object that can be reused. For now, you'll proceed with the stored credentials and you will use them in a later step..

3. You're going to use an array of hostnames to connect to individual ESXi hosts for configuration. To create the array, you set a variable and store a comma separated list of addresses to connect to. The addresses can either be hostnames or IP addresses. For this example, you will use IP addresses, but they can easily be fully qualified domain names:

```
# Step 2 - Set a list of target ESXi hosts and IP settings
$esxiTargets = "192.168.0.241","192.168.0.242", "192.168.0.243",
"192.168.0.244"
```

4. For the network configuration settings, you will need to set up some additional settings. Since each host has three additional vmkernel ports configured, you need to build a different address for each of these to be used in Step 8. To allow this, you will create three additional variables that contain the first three octets of the network for each vmkernel port:

```
$vMotionNetwork = "192.168.50."

$storageNetwork = "192.168.100."

$ftlogNetwork = "192.168.200."
```

5. The next step is to go back and pull in all of the code you had previously written in one form or another. For this, you will reuse the ForEach loop to execute the cmdlets on multiple ESXi hosts:

```
# Step 3 - Write a ForEach loop to iterate through hosts
ForEach ($hostname in $esxiTargets) {
```

6. The curly brace marks the beginning of the ForEach loop. You will close the loop with a right curly brace later in the script. Inside the loop, you're going to include Steps 4 – 9 from the outline.

7. For the next step, you're going to use our stored credentials to connect to an ESXi host. Immediately after this, you will store our VMHost object for use throughout the rest of the loop:

```
# Step 4 - Connect to ESXi host
$connectedHost = connect-viserver $hostname -Credential $esxiCreds


# Step 5 – Obtain a VMHost object to use for configuration
$esxihost = Get-VMHost $hostname
```

8. For the next several steps, you're just going to pull code you have already developed. Since each step was covered in depth, you will just bring over the code:

```
# Step 6 – Join the ESXi system to Active Directory
$esxihost | Get-VMHostAuthentication |
```

```
Set-VMHostAuthentication -JoinDomain -Domain domain.local -user
username -password ***** -Confirm:$false


# Step 7 - Enable services on the ESXi host & set firewall
$esxihost | Get-VMHostService | where { $_.key -eq "TSM-SSH" } |
Set-VMHostService -Policy "On" -Confirm:$false


$esxihost | Get-VMHostService | where { $_.key -eq "TSM-SSH" } |
Start-VMHostService -Confirm:$false


# Step 8 - Configure the network settings
$esxihost | Get-VirtualPortGroup -Name "VM Network" | Remove-
VirtualPortGroup -Confirm:$false
```

9. For the network settings, you will need three additional IP addresses for the vMotion, Storage, and FT Logging vmkernel ports. You will compute these addresses using the last octet of the service console IP. To do this, you will first retrieve the IP address of the host:

    ```
    $esxihost_ipaddress = $esxihost | Get-VMHostNetworkAdapter -name
    vmk0
    ```

10. Next, you will split the string based on the period between octets, then take the last octet of the IP address and store it as a variable. The IP is in a property called `IP`. To split that IP into an array, you will use the `Split()` method, which is a built-in PowerShell method that transforms a string into an array by separating characters with the character passed into the method.

    For instance, you want to separate the string at the periods of the IP address, so you pass `"."` into the `Split()` method. Since the `Split()` method turns it into an array, you can then reference the element you want to return – the fourth element. However, remember arrays begin count at 0, so you will return element 3 using square brackets.

    ```
    $lastOctet = $esxihost_ipaddress.IP.Split(".")[3]
    ```

> Because data is stored in objects, objects have both properties and methods. Methods perform operations on the data of the object and properties contain the data of the object. In subsequent recipes throughout this book, you will look at and use other methods to gain more experience using built-in PowerShell functionality to manipulate data stored in objects.

11. The last step to build the address for this host in the `ForEach` loop is to concatenate the final octet with the network strings to build a full IP address:

    ```
    $vmotionIP = $vMotionNetwork + $lastOctet

    $storageIP = $storageNetwork + $lastOctet

    $ftlogIP = $ftlogNetwork + $lastOctet
    ```

12. Now that your unique IP addresses are created on the three additional networks, you can use them with the cmdlets you wrote in the *Setting network configuration* recipe.

```
$esxihost | New-VMHostNetworkAdapter -PortGroup "VMotion Network"
-VirtualSwitch vSwitch0 -IP $vmotionIP -SubnetMask 255.255.255.0
-VMotionEnabled $true


$esxihost | Get-VirtualPortGroup -Name "VMotion Network" | Set-
VirtualPortGroup -VlanID 50


# Create new virtual switch for Storage and FT Logging
$esxihost | New-VirtualSwitch -Name vSwitch1 -Nic vmnic2,vmnic3


# Create vmkernel ports for Storage and FT Logging
$esxihost | New-VMHostNetworkAdapter -PortGroup "Storage Network"
-VirtualSwitch vSwitch1 -IP $storageIP -SubnetMask 255.255.255.0
-VsanTrafficEnabled $true


$esxihost | Get-VirtualPortGroup -Name "Storage Network" | Set-
VirtualPortGroup -VlanID 100


$esxihost | New-VMHostNetworkAdapter -PortGroup "FT Logging
Network" -VirtualSwitch vSwitch1 -IP $ftlogIP -SubnetMask
255.255.255.0 -FaultToleranceLoggingEnabled $true


$esxihost | Get-VirtualPortGroup -Name "FT Logging Network" | Set-
VirtualPortGroup -VlanID 200


# Create new Virtual Switch for Virtual Machines
$esxihost | New-VirtualSwitch -Name vSwitch2 -Nic vmnic4,vmnic5


# Create Port Groups for Virtual Machines
New-VirtualPortGroup -Name "Infrastructure Network"
-VirtualSwitch vSwitch2 -VLanId 1 -Server $connectedhost


New-VirtualPortGroup -Name "Application Network"
-VirtualSwitch vSwitch2 -VLanId 2 -Server $connectedhost


# Step 9 - Configure NFS & iSCSI settings
# Connect NFS datastore
```

```
$esxihost | New-Datastore -Nfs -Name DataStoreName -Path /data1/
export -NfsHost nfsserver.domain.local


# Configure iSCSI Settings

$esxihost | Get-VMHostStorage | Set-VMHostStorage
-SoftwareIScsiEnabled $true$iSCSIhba = $esxihost | Get-VMHostHba
-Type iScsi


New-IScsiHbaTarget -IScsiHba $iSCSIhba -Address $target
-ChapType Required -ChapName vSphere -ChapPassword Password1


$esxcli = Get-ESXCLI -VMHost $esxihost


$esxcli.iscsi.networkportal.add($iscsihba, $true,"vmk2")
```

13. The final part of the ESXi host configuration is closing the `ForEach` loop and then disconnecting from this host so that you can connect to the next host:

    ```
    Disconnect-VIServer -Server $connectedHost -Confirm:$false
    }
    ```

    At this point in the initial configuration, you would want to format your datastores on iSCSI or Fibre Channel arrays, but this is not really a repeatable set of steps. I would suggest one of the two things—either configure the datastore manually from PowerCLI or configure it from the GUI and then come back and run the remainder of the script. Since the focus of this example is to make a repeatable configuration script, the datastore formatting doesn't fit since it is a command used just one time .

14. The next step is to take our hosts and connect them to vCenter. The easiest way to do this is to connect to vCenter and then use `Add-VMHost` to add them into inventory. While in the same `ForEach` loop to accomplish this, you can set central syslog and rescan the hosts for all storage changes:

    ```
    $vcenter = connect-viserver vcentersrv.domain.local

    $datacenter = Get-Datacenter "Primary"
    ```

> For the purpose of this script, you are going to assume that vCenter already has a datacenter created and named "Primary." You will use this location to place the ESXi host into vCenter.

15. Next, you will run through an additional `ForEach` loop to add the hosts and set their settings in vCenter:

    ```
    ForEach ($hostname in $esxTarget) {
    ```

16. Now, you are ready to add the host into vCenter from the *Joining an ESXi host to vCenter* recipe:

```
# Step 10 - Join hosts to vCenter

Add-VMHost -Server $vcenter -Name $hostname 1 -Location
$datacenter -Credential $esxiCreds
```

17. After adding the host to vCenter, you want to store a `VMHost` object pointing to the host to use with later cmdlets in this loop:

```
$esxihost = Get-VMHost $hostname
```

18. For the next few steps, you will pull the host settings related to rescanning for datastores and setting the syslog settings:

```
# Step 11 - Rescan for storage changes

$esxihost | Get-VMHostStorage -RescanAllHBA -RescanVmfs


# Step 12 - Configure persistent syslog storage

$esxihost | Get-AdvancedSetting | Where {$_.Name -like "Syslog.
global.logDirUnique"} | Set-AdvancedSetting -value $true
-Confirm:$false


$esxhost | Get-AdvancedSetting | Where {$_.Name -like "logDir"}
| Set-AdvancedSetting -value "[iSCSIDatastore1] syslog"
-Confirm:$false
```

19. Finally, you will close the loop with a right curly brace.

```
}
```

> With `connect-viserver`, you might have to log in a second time in the script with different credentials to vCenter versus individual hosts. Afterwards, the hosts should be populated into vCenter.

Finally, your settings and desired state should be fully transferred to the ESXi host by the script.

How it works...

In this example, you wrap up all of the code you have developed throughout the chapter. You bring together the pieces of code that achieve specific tasks into a fully scripted configuration that you can apply toward a number of ESXi hosts. The script gives us repeatability, so when you need to extend the cluster with a new ESXi, or when you rebuild the host because you've replaced or upgraded the hardware, you can run this script against it and be back to the same working condition as before replacement.

The basis of the script is a `ForEach` loop. Because you define the ESXi hosts in an array, you can connect to each of them and run all of the commands and then move the next entry in the array. The script also suppresses confirmation dialogs so that it can continue to issue cmdlets against the host. You also stored the login credentials, which means that you only have to log in once and the script will use the same credentials to connect and configure all of the hosts in the defined array.

See also

- VMware vSphere Host Profiles at `http://www.vmware.com/products/vsphere/features/host-profiles`

2

Configuring vCenter and Computing Clusters

In this chapter, you will cover the following topics:

- ▶ Creating a virtual datacenter in vCenter
- ▶ Creating a cluster and adding ESXi hosts
- ▶ Setting cluster advanced features, including HA, DRS, and EVC
- ▶ Setting up resource pools
- ▶ Setting up folders to organize objects in vCenter
- ▶ Setting permissions on vCenter objects

Introduction

A single ESXi host allows you to run multiple virtual machines on a single server, but to tap the full potential of power from vSphere, you're going to need vCenter and clusters of ESXi hosts. This chapter will cover the basic concepts of creating and managing pools of resources using vCenter and multiple ESXi hosts.

vCenter is an increasingly critical part of the vSphere infrastructure since it handles the coordination of clustering and automation across multiple ESXi hosts. This drives the change and increases the complexity of vCenter deployments in each new version of vSphere. Even while the vCenter deployments are becoming more complex, VMware is working to try and ease that management by packaging the solution in simpler ways. vSphere 5.1 introduced the new **Single Sign-On** (**SSO**) service to the platform, and vSphere 5.5 streamlined deployment of the SSO's second version in vSphere. vSphere 5.5 also improved the virtual appliance version of vCenter, known as the **vCenter Server Appliance (VCSA)**. With version 6.0, the SSO service has grown into the **Platform Services Controller** (**PSC**). In addition to SSO, the PSC includes licensing, a certificate authority, and a centralized certificate store. The PSC can also replicate data between multiple instances of itself.

For the purpose of this chapter, you assume that vCenter is set up and it is in an operational state. If you do not already have vCenter running, deploying VCSA as a virtual appliance is, by far, the easiest way to get it running and functional for your environment.

For deploying VCSA for version 5.5, following the prompts in the **Deploy OVF Template...** menu option of the GUI is the easiest way to deploy vCenter. The deployment of vCenter is not within the scope of this book since it really requires the GUI to deploy. Even though PowerCLI can deploy virtual appliances, it misses answering the critical questions needed for the successful VCSA deployment.

If you are deploying VCSA for version 6.0 and you are using VMware Workstation, VMware Fusion, or even standalone ESXi to deploy your VCSA, using the VMware OVF Tool is a quick and automated method. OVF Tool allows you to define the passwords, IP addresses, and other information needed for VCSA to perform its configuration during the first boot. Without these parameters, the VCSA fails to configure on first boot and you receive an error. Blogger William Lam has a post and scripted installation using the OVF Tool at `http://www.virtuallyghetto.com/2015/02/ultimate-automation-guide-to-deploying-vcsa-6-0-part-1-embedded-node.html`.

Creating a virtual datacenter in vCenter

vSphere has several defined objects that are used to create virtual datacenters. For example, an object named `Datacenter` sits at the root of vSphere and allows the clusters and other host infrastructures to be placed inside the virtual datacenter. Installations might have multiple datacenters; however, most VMware administrators use additional datacenter objects in vSphere to represent a physical site and use the datacenter object as a boundary where the infrastructure exists. In this recipe, you will take a look at the code needed to create your new datacenter object in vCenter.

Getting ready

For this recipe, you will need to open a PowerCLI prompt, you need the DNS name or IP address of your vCenter host, and the password for the default administrator account in vCenter.

How to do it...

In order to create a virtual datacenter, and to create new datacenter object in vCenter, perform the following steps:

1. The first step is to connect to your vCenter server. You need to use the same cmdlet to connect to vCenter that you use to connect to a single ESXi host: the `Connect-VIServer` cmdlet:

   ```
   Connect-VIServer vcentersrv.domain.local
   ```

2. Log in with the default `Administrator@vsphere.local` account created during the VCSA deployment or vSphere SSO installation. When you log in successfully, a prompt will be displayed that shows the server you are successfully connected to, as shown in the following screenshot:

3. To start, let's run the `Get-Datacenter` cmdlet with no additional parameters to see whether there are any existing objects in vCenter. On a fresh vCenter install, there is no output:

```
Get-Datacenter
```

4. To configure a new datacenter, the cmdlet is `New-Datacenter`, which is very straightforward. Logically, you just need to provide a name for your datacenter, and it will create a datacenter on the vCenter:

```
New-Datacenter -Name "Primary"
```

5. If you run the preceding cmdlet, you'll receive an error that a mandatory parameter and location is missing. However, if this is a brand new vCenter installation, which location would you possibly pass into this cmdlet?

6. To answer this question, you will run the `Get-Folder` cmdlet to see whether there are any folder locations that you might be able to use. You will add a `-NoRecursion` parameter because you only want to return the top-level results:

```
Get-Folder -NoRecursion


Name                         Type

----                         ----

Datacenters                  Datacenter
```

7. Interestingly, there is a root folder called `Datacenters` that exists by default. That's a location you can pass into this cmdlet.

8. So, the next step is to put the two cmdlets together and create our datacenter. You will repeat the `New-Datacenter` cmdlet and you will specify `-Location` this time with the cmdlet in Step 4 returning the `Datacenters` folder:

    ```
    New-Datacenter -Name "Primary" -Location (Get-Folder -NoRecursion)


    Name

    ----

    Primary
    ```

9. The resultant output confirms that a new datacenter called `Primary` has been created. If you rerun the `Get-Datacenter` cmdlet, now, it has the same output.

How it works...

The cmdlet that actually creates a new datacenter is very straightforward. The only point of confusion is a required `-Location` parameter. This requirement is confusing because there are no objects in vCenter on a fresh install. However, as you explore the `Get-Folder` cmdlet, you will find that a default `Datacenters` folder is created during the installation of vCenter and it is meant to hold new datacenter objects. The following diagram depicts the hierarchy:

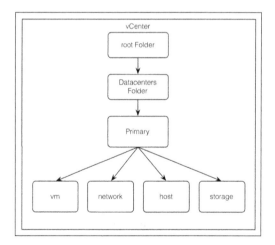

At the top level is the root folder of vCenter. Inside the root folder is the `Datacenters` folder. Before a datacenter is created, root is the only folder that exists in vCenter. The new `Primary` datacenter that you created is located inside the `Datacenters` folder. Inside `Primary`, four additional folders are automatically created that correspond to the four views that you see in the vSphere Client. Each of these are special folders used by vCenter services to house the inventory items.

By passing location in the `Datacenters` folder using `Get-Folder -NoRecursion`, you know that you are passing location in the root folder where our datacenter named `Primary` should be created. If you rerun `Get-Folder | Select *` after creating the datacenter, you will see the additional objects in the preceding figure, and you will see that their `Parent` parameter is defined as `Primary`.

Creating a cluster and adding ESXi hosts

Clusters are the basis for everything that is great within vSphere. Clusters are the level where individual resources become pooled and shared for virtual machines. Clusters allow all higher-level functionalities within vSphere, such as an automatic restart after a hardware failure and dynamic balancing of workloads. Individual ESXi hosts and clusters can exist at the same level under a datacenter object in vSphere.

In this recipe, you will walk through the steps necessary to set up your first cluster of servers in vCenter. You will be reusing the same four ESXi hosts that you configured in the *Creating a configuration script to set all properties uniformly* recipe from *Chapter 1, Configuring the Basic Settings of an ESXi Host with PowerCLI*.

Getting ready

This chapter assumes that your vCenter has the datacenter object defined and that you have individual ESXi hosts connected or managed by vCenter. In this example, you will call the `Primary` datacenter, like you defined in the previous recipe. You will need to open a PowerCLI prompt and connect to your vCenter instance.

vSphere clusters have several advanced features, and when creating the cluster, you will want to enable these features in most cases. In this example, you will enable **High Availability** (**HA**) and **Dynamic Resource Scheduling**. You will also set the DRS mode to `FullyAutomated`. In the next recipe, you will look at configuring these cluster settings.

How to do it...

In order to set up your first cluster of servers in vCenter by reusing the same four ESXi hosts that you have configured earlier, perform the following steps:

1. The `New-Cluster` cmdlet is very simple. If you perform a `Get-Help` cmdlet on this cmdlet, you will see a number of additional parameters that you can define, but in its simplest form, `New-Cluster` only requires two parameters: a location and a name:

   ```
   New-Cluster -Location (Get-Datacenter -Name "Primary")
   -Name BigCluster -HAEnabled -DRSEnabled -DRSAutomationLevel
   FullyAutomated
   ```

2. A confirmation output will follow and you can confirm whether the new cluster was created using the `Get-Cluster` cmdlet.

3. The next step is to add a host to a cluster. For this, you will use the familiar `Get-VMHost` cmdlet to find our existing host and then use the `Move-VMHost` cmdlet to move it into the cluster, passing the cluster object as the location:

```
Get-VMHost esxsrv1.domain.local | Move-VMHost -Destination (Get-Cluster -Name "BigCluster")
```

4. If you had not previously added the host to vCenter, you could simply use the `Add-VMHost` cmdlet that you used in the *Joining an ESXi host to vCenter* recipe in *Chapter 1, Configuring the Basic Settings of an ESXi Host with PowerCLI*, and specify `-Location` to be our newly created cluster. In vCenter Server 6, you also need to add the `-Force:$true` parameter. vCenter Server 6 checks the SSL thumbprint of the host and if it's not trusted, the `Add-Host` cmdlet will fail. The `-Force` parameter will make this host add, even with an untrusted certificate.

```
Add-VMHost -Name esxsrv1.domain.local -Location (Get-Cluster "BigCluster") -Force:$true
```

How it works...

Creating a cluster on vSphere is pretty simple. A cluster can exist without any hosts in it, but there's no reason to set up empty clusters. Adding hosts to the cluster begins to build a functional pool of resources for your virtual machines to share.

Creating a cluster only needs a location and a name. Everything beyond these parameters is optional, but you can get much more detailed information if you like by enabling and configuring advanced features, such as DRS and HA, from the same cmdlet.

There's more...

In the previous chapter, we covered how to add an individual host to vCenter and configure multiple hosts with the same configuration. This recipe takes a step further and allows you to create clusters using individual ESXi hosts with similar configurations.

Admitting the first host into a cluster is simple from PowerCLI, but it certainly has implications for the new cluster. The first host of the cluster defines some things about the cluster and determines what can be added in the future. Hosts should contain compatible processors or should have other settings configured to make disliked processors from the same vendor more compatible with each other. These settings are EVC settings and we will cover these in the next recipe.

See also

▸ The *Joining an ESXi host to vCenter* and *Creating a configuration script to set all properties uniformly* recipes in *Chapter 1, Configuring the Basic Settings of an ESXi Host with PowerCLI*

Setting cluster advanced features, including HA, DRS, and EVC

The previous recipe mentioned the advanced features of vSphere clusters. In this recipe, you will configure those advanced features since it is more common to reconfigure these settings than to initially set these settings.

The cluster settings you are going to be working with are HA, DRS, and **Enhanced vMotion Compatibility** (**EVC**) settings. In the vSphere Client, these settings are exposed in the **Edit Cluster Settings** option.

Creating a cluster is a one-time event, but as you deploy vSphere, you might not be ready to automate vMotions or Storage vMotions with DRS in your cluster from the beginning.

However, over a period of time, your comfort level with these automation technologies begins to increase and you would want to put the cluster on autopilot and change the automation level to be fully automated. You will cover how to do this and how to set up other common settings from PowerCLI.

PowerCLI is useful for these settings because it allows you to repeat the same cmdlet against multiple clusters in large environments, or to change your configuration and change it back easily for smaller environments. However, even cluster-wide settings are just the beginning of what you can configure faster in PowerCLI than in the GUI.

DRS rules are a great example of something that is faster to configure from PowerCLI than in the GUI. Since you can use the `Get-VM` cmdlet to quickly return an object with multiple VMs matching a search string, you can pass this into a new rule instead of searching and clicking multiple times on the GUI.

You can also see that one vSphere advanced clustering feature is missing from this recipe: **Fault Tolerance** (**FT**). The reason it is missing from this configuration recipe is that the only requirements for Fault Tolerance to work were already configured in the ESXi configuration from *Chapter 1, Configuring the Basic Settings of an ESXi Host with PowerCLI*. For FT, you need to set a vmkernel port with an IP address and enable this for FT logging. There is no additional configuration required.

Getting ready

To begin, you will need to open a PowerCLI window and log into a vCenter server.

In this recipe, you will configure the availability and resource balancing features of a cluster. All of these features are managed at a cluster level, so you will utilize the `Get-Cluster` cmdlet to specify which cluster you want to be working with.

Reconfiguring the cluster settings is a common requirement for the existing clusters. You will explore simple cmdlets to enable and disable HA or DRS on a cluster, you will take a look at how to configure the additional settings used for the restart order with HA and how workloads are balanced with DRS.

However, DRS doesn't stop at simply balancing workloads. DRS rules expand beyond simply spreading the load evenly across the hosts in a cluster. DRS rules can dictate which VMs should coreside on the same host and which VMs should never reside on the same host. The latter is particularly helpful when you have multiple, identical app servers fronted by a load balancer. To have redundancy, you need to make sure that a hardware failure can't take down both VMs at once. You will also examine how to create simple DRS rules for keeping VMs together and keeping VMs separated.

For HA, **Admission Control** is a feature that reserves resources, so that a cluster can withstand losing one or more hosts without negatively impacting the performance. This setting can be enabled or disabled, and you can also adjust the number of hosts' failures that the cluster can accommodate. The cluster reserves resources so that if the specified numbers of hosts fail, the VM workloads still run without being resource constrained. These settings prevent users from powering on virtual machines if the resource conditions are not met.

 If you are running vCloud Director, DRS is required and must not be disabled.

Since HA is primarily concerned with recovering failed virtual machines, there are two additional settings that you can set: **Isolation Response** and **Restart Priority**. Isolation Response sets the reactive behavior that the host should take if it becomes isolated from the rest of the cluster. You can set this to either power off the VMs or to *do nothing*. The Restart Priority setting sets the default priority for VMs in the cluster if they fail. This can be set per VM, so at the cluster level you are setting the default. Ideally, your most critical VMs are manually set to a higher level, other critical VMs to medium, and management systems and noncritical systems to low. It's also important to note that if Admission Control doesn't have resources to restart any more VM's, the lowest priority VMs would be left powered off.

How to do it...

In order to set up the advanced features of a cluster, including HA, DRS, and EVC, that is, to configure the availability and resource balancing features of a cluster, perform the following steps:

1. Changes to any existing cluster will utilize the `Set-Cluster` cmdlet. The `Set-Cluster` cmdlet has the same features as the `New-Cluster` cmdlet you used in the previous recipe. With `Set-Cluster`, you will specify a cluster using the `-Cluster` parameter and then you can make any configuration changes to the cluster you want. Let's start with a quick cmdlet to disable HA:

   ```
   Set-Cluster -Cluster "BigCluster" -HAEnable $false
   ```

   Change `$false` to `$true` and the cmdlet turns on the feature. Simple!

   ```
   Set-Cluster -Cluster "BigCluster" -HAEnable $true
   ```

2. Next, you might want to change the **Admission Control** and **Failover Level** settings for HA on the cluster. Again, you turn to the `Set-Cluster` cmdlet to make these setting changes. The `-HAAdmissionControlEnabled` parameter controls whether Admission Control is turned on. The `-HAFailoverLevel` parameter is set to a number from 1 to 4 specifying how many host failures you want the cluster to be able to survive. You will set our example to survive one host failure:

   ```
   Set-Cluster -Cluster "BigCluster"
   -HAadmissionControlEnabled $true -HAFailoverLevel 1
   ```

3. Next, you can set the Isolation Response and Restart Priority settings for the cluster, again using the `Set-Cluster` cmdlet. First, you use `-HAIsolationResponse` to set the behavior if the host becomes isolated. Next, you use `-HARestartPriority` to set the default priority to restart VMs in the cluster:

   ```
   Set-Cluster -Cluster "BigCluster" -HAIsolationResponse PowerOff
   -HaRestartPriority Medium
   ```

   It is also important to note that all of these settings can be combined in a single `Set-Cluster` cmdlet.

4. It is also common to change the DRS mode on a cluster. To do this, you again use the `Set-Cluster` cmdlet, but you will use the `-DrsAutomationLevel` parameter to set the mode:

   ```
   Set-Cluster -Cluster "BigCluster" -DrsAutomationLevel Manual
   -Confirm:$false
   ```

   More commonly, you might want to set the DRS mode to fully automated:

   ```
   Set-Cluster -Cluster "BigCluster" -DrsAutomationLevel
   FullyAutomated -Confirm:$false
   ```

5. Next, in this example, you will make sure that our domain controllers are not both running on the same ESXi node by defining a DRS rule. First, you need to retrieve a list of the domain controller VM's with the `Get-VM` cmdlet. The `New-DrsRule` cmdlet allows you to create a KeepTogether or a Separate rule. The syntax is very simple. You need to specify a name for our rule, a cluster, whether or not this is a KeepTogether rule, and finally, which VMs are passed by a variable:

```
$domaincontrollers = Get-VM -Name "DC*"


New-DrsRule -Name "Separate DCs" -Cluster "BigCluster"
-Enabled $true -KeepTogether $false -VM $domaincontrollers
```

6. Reporting the EVC mode setting is very straightforward from PowerCLI. To begin, you must retrieve the cluster object and the EVC mode setting as a parameter of this object:

```
Get-Cluster "BigCluster" | Select Name, EVCMode
```

7. Changing the EVC mode setting is pretty simple using the `Set-Cluster` cmdlet, but it requires PowerCLI 5.5 R2. The earlier versions of PowerCLI do not include the `-EVCMode` parameter with the `Set-Cluster` cmdlet. You simply need to specify the key of the mode you want to enable. A chart of the keys is included in the *How it works...* section:

```
Set-Cluster -Cluster "Primary" -EVCMode 'intel-penryn'
```

How it works...

All of these settings are cluster-level settings and so most of them are set using the `Set-Cluster` cmdlet. `Set-Cluster`, like `New-Cluster` in the previous recipe, has a number of parameters specific to vSphere clusters. Many of the parameters are simple Boolean inputs, either `$true` or `$false`. Others have defined the input where you might need to use the `Get-Help Set-Cluster` cmdlet in order to investigate the exact input expected for the setting you desire.

Again, where does PowerCLI buy you the biggest benefits for cluster configuration? Anytime you've got more than one cluster, and you want to make sure that you have uniform settings across the clusters. If you want to make sure that your HA, DRS, and EVC settings are the same for several hosts, you simply use a `Get-Cluster` cmdlet. Search for all of the clusters in your datacenter or for specific clusters in the datacenter, and then pipe that directly into a `Set-Cluster` cmdlet with your desired settings. This is much faster and helps you to eliminate human error by picking the wrong setting in the GUI while changing clusters one at a time.

One of the parameters, `-EVCmode`, specifically needs enumerated input settings. The following tables show the available EVC mode keys that can be set from the `New-Cluster` and `Set-Cluster` cmdlets:

Intel CPUs	EVC mode keys
Intel® "Merom" Generation	intel-merom
Intel® "Penryn" Generation	intel-penryn
Intel® "Nehalem" Generation	intel-nehalem
Intel® "Westmere" Generation	intel-westmere
Intel® "Sandy Bridge" Generation	intel-sandybridge
Intel® "Ivy Bridge" Generation	intel-ivybridge

AMD CPUs	EVC mode keys
AMD Opteron™ Generation 1	amd-rev-e
AMD Opteron™ Generation 2	amd-rev-f
AMD Opteron™ Gen. 3 (no 3DNow!™)	amd-greyhound-no3dnow
AMD Opteron™ Generation3	amd-greyhound
AMD Opteron™ Generation 4	amd-bulldozer
AMD Opteron™ "Piledrive" Generation	amd-piledriver

There's more...

While you talked about KeepTogether and Separate DRS rules, there are other types of DRS rules and those are VM to Host rules. While PowerCLI doesn't provide cmdlets to handle DRS Affinity Group assignments, this is one of the best use cases for PowerCLI and one that I use in my managed environments. Users who have implemented a VMware Metro Storage Cluster might need to routinely manage and assign VMs to Affinity Groups to make sure that VMs are running in a particular physical location. This is an advanced use case, but you should definitely read the blog post and code from Niklas Åkerlund linked in the *See also* section.

See also

▸ Refer to the article titled, **vSphere Cluster Host-VM rule affinity with PowerCLI**, by Niklas Åkerlund, available at `http://vniklas.djungeln.se/2012/06/28/vsphere-cluster-host-vm-rule-affinity-with-powercli/`

Setting up resource pools

Resource pools are objects within vSphere where VM objects with similar performance requirements can be grouped together. Resource pools allow a priority to be set to pools of compute, memory, and disk resources, so that when the contention occurs, the hypervisor can choose which VMs get access to resources first. Resource pools exist within the cluster objects in vSphere or within host objects if a host is not a part of a cluster.

Getting ready

To begin this recipe, you will need to open a PowerCLI window and connect to a vCenter server. For the purpose of this example, we're going to configure two resource pools: `Production` and `Development`. The `Production` resource pool will be configured with the `high` setting for CPU and memory resources. The `Development` resource pool will be configured with the `low` setting for CPU and memory resources.

Sometimes, PowerCLI defines its normal conventions and resource pools is one of them. You can't simply pass a cluster in as the location for a new resource pool. Although it is not shown, when you create a cluster, it creates a default resource pool called `Resources`, which is similar to how the creation of a datacenter created four subfolders. The location that a `New-ResourcePool` cmdlet is looking for is a resource pool object.

How to do it...

In order to set up a resource pool, perform the following steps:

1. The first step is to locate the root `Resources` folder so that you can use it in the creation of a new resource pool. Since all of the root resource folders are called `Resources`, you should scope the `Get-ResourcePool` cmdlet to make sure that the pool is for the correct cluster. If you only have one cluster, this is a nonissue, but you will illustrate it to make the code more reusable:

   ```
   Get-ResourcePool -Name "Resources" -Location (Get-Cluster -Name
   "BigCluster")
   ```

2. With this scoping statement, you can use this in the `-Location` parameter of the `New-ResourcePool` cmdlet. In addition to the location, you also need to specify a name for the new pool and additional parameters to define the CPU and RAM share settings. Additional parameters can also be defined to set reservations for CPU or RAM and expandable reservations. In our example, you will set the `-CPUExpandableReservations` and `-MemExpandableReservation` parameters to `$true`:

   ```
   New-ResourcePool -Name "Production" -Location (Get-
   ResourcePool -Name "Resources" -Location (Get-Cluster -Name
   "BigCluster") ) -CPUSharesLevel high -MemSharesLevel high
   -CpuExpandableReservation $true -MemExpandableReservation $true
   ```

3. While this is the most correct way to create the new pool, you need to ensure that you have specified the correct resource pool to contain it. There is a much shorter cmdlet that will accomplish the same in our environment:

```
New-ResourcePool -Name "Production" -Location
"BigCluster" -CPUSharesLevel high -MemSharesLevel high
-CpuExpandableReservation $true -MemExpandableReservation $true
```

This is much cleaner and more readable code than the previous one and it will accomplish the same thing.

4. The next step is to repeat the same code for our Development resource pool, except you want to set the share levels to low in this example:

```
New-ResourcePool -Name "Development" -Location "BigCluster"
-CPUSharesLevel low -MemSharesLevel low -CpuExpandableReservation
$true -MemExpandableReservation $true
```

5. Again, you have to move objects into this resource pool. Moving a VM into the resource pool will not move it out of the folders or other locations where it might be assigned, it will only move the VM in the context of the Host and Clusters view. You will use the Move-VM cmdlet and specify the host and the location:

```
Move-VM -Name vCenterSrv -Location (Get-ResourcePool "Production")
```

6. Lastly, if you have an existing resource pool, but you need to adjust the settings, you can do so with the same parameters using the Set-ResourcePool cmdlet:

```
Set-ResourcePool -ResourcePool (Get-ResourcePool Production)
-CpuSharesLevel Custom -NumCpuShare 8000
```

How it works...

The New-ResourcePool cmdlet creates a new pool inside the location specified in the cmdlet. If the location is a Host or Cluster, the new pool is automatically placed into the Resources pool at the root of the cluster or host.

The New-ResourcePool cmdlet provides a lot of additional parameters to configure the resource pool from the start. In our example, you specified the shares level, which is one of four enumerated choices: Low, Normal, High, and Custom. With Custom, you also have to specify a number using the -NumCpuShares and -NumMemShares parameters. In addition to share definitions, you can also set the reservations and limits for CPU and RAM. You can specify a number of MHz or MB for reservations and limits on the pool. Limits allow no more than the specified amount of CPU or RAM resources, and reservations guarantee the specified amount of CPU or RAM resources for the pool. There is also the concept of expandable reservations that allows a pool to borrow the specified value if its parent has unallocated resources.

One thing that should start to become clear is that unique names go a long way to shortcutting your code. If a name is unique to a single folder, cluster, host, or an object, there is no need to pass in the location by an object using a `Get-` cmdlet. Names without spaces also help to shortcut code, since any name with spaces requires quotes around it.

There's more...

Resource Pools in vSphere can be intimidating, but they play a powerful role in keeping things running smoothly. Chris Wahl of WahlNetwork.com has an excellent post about Resource Pools and includes a PowerCLI script to help keep your pools balanced using his formula for computing the appropriate number of shares. For more information, refer to `http://wahlnetwork.com/2012/02/01/understanding-resource-pools-in-vmware-vsphere/`

See also

- ▸ Creating and reporting vSphere resource pools
- ▸ Moving objects between resource pools
- ▸ Reporting shares, reservations and limits of resource pools, and virtual machines
- ▸ Setting shares, reservations, and limits for similarly classified objects in vSphere

Setting up folders to organize objects in vCenter

vSphere folders are flexible containers with other vSphere objects inside. Folder objects in vSphere are meant to be a logical organizational structure for objects that are not tied to physical resources. This means that VM objects from different clusters or even different datacenters can be logically grouped together. The same applies to port groups, switches, or datastores.

This is important as you begin to look at delegating access from VMs to operators, developers, and other users in the organization, so that you can group together all of the VMs that a user needs to access. Folders also help administrators to easily locate objects and report on objects for a particular business unit or group within their companies.

In this recipe, you will look at the simple cmdlets used to create folder structures in vSphere, and move objects into these folders with simple PowerCLI cmdlets.

Getting ready

To begin, you need to open a PowerCLI window and connect to a vCenter server. You should also read and review the *Creating a virtual datacenter in vCenter* recipe earlier in this chapter, since it discusses the hierarchy of folders inside a vSphere datacenter. This recipe uses a lot of the concepts introduced in the earlier recipes.

For this recipe, you will use the `New-Folder` cmdlet and understand the different types of folders that it can create inside vCenter. Folders are used in multiple areas of vCenter for organizational purposes. You will also take a look at the use of `Get-Folder` and `Remove-Folder`.

The `New-Folder` cmdlet is another cmdlet that relies on the `-Location` parameter to determine where to create the object you're defining. As you observed in the *Creating a virtual datacenter in vCenter* recipe, you are able to use the `Get-Folder` cmdlet to return the four special folders automatically provisioned under a datacenter object.

For this example, you will create several folder structures. You will create two, two-level folders under the **VM and Templates** view for **Infrastructure** and **App Servers**. You will create two subfolders called **Domain Controllers** and **VMware** under **Infrastructure**. You will create a **Standard vSwitches** folder in the **Networks** view and you will create an **NFS** and **iSCSI** folder under the **Datastores** view. Finally, you will create a **Finance** and **IT** folder under the **Host and Clusters** view to store clusters owned by these businesses. The following illustrates this structure:

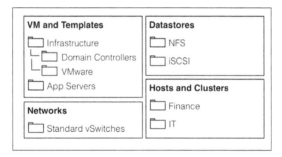

How to do it...

In order to set up folders to organize objects in vCenter, perform the following steps:

1. The first step is to retrieve the datacenter where you want these folders to be created. By first getting the datacenter object, you ensure that if you had more than one datacenter defined, you would be operating in the correct datacenter. So, the first step is to run `Get-Datacenter` and find our `Primary` datacenter:

    ```
    Get-Datacenter -Name "Primary"
    ```

In this particular example, since `Primary` is the only datacenter object, you do not have to pass in the datacenter object to the next cmdlet. By specifying `Primary`, you ensure that you have selected the desired datacenter if you have multiple objects with the same name in your infrastructure.

2. The next step is to pipe this datacenter object into a `New-Folder` cmdlet, which will limit the results to within this datacenter. If you start with the `Infrastructure` folder, you need to return the root folder with type `vm`:

```
Get-Datacenter -Name "Primary" | Get-Folder -name "vm"
```

3. Now that you have a single folder, this will serve as our location parameter for our new `Infrastructure` folder. Using `New-Folder`, you will pass in our desired name and the location parameter from the previous step:

```
New-Folder -Name "Infrastructure" -Location (Get-Datacenter -Name
"Primary" | Get-Folder -name "vm")
```

4. Next, you can repeat the same step with our `App Servers` folder. The only parameter that should change is the name parameter:

```
New-Folder -Name "App Servers" -Location (Get-Datacenter -Name
"Primary" | Get-Folder -name "vm")
```

5. The next step is to create a subfolder under Infrastructure for Domain Controllers. To do this, you change the name and the location of the same cmdlet. Instead of searching for the folder named `vm`, you will search for the one you just created named `Infrastructure`. While we're at it, you can create the `VMware` folder as well by repeating the cmdlet with a different name defined:

```
New-Folder -Name "Domain Controllers" -Location (Get-Datacenter
-Name "Primary" | Get-Folder -name "Infrastructure")


New-Folder -Name "VMware" -Location (Get-Datacenter -Name
"Primary" | Get-Folder -name "Infrastructure")
```

6. Next, you will move to create the `Standard vSwitches` folder under the Networking area. To do this, you need to run the `New-Folder` cmdlet and search for the root `network` folder in the datacenter for the location:

```
New-Folder -Name "Standard vSwitches" -Location (Get-Datacenter
-Name "Primary" | Get-Folder -Name "network")
```

7. With this successfully created, you will write three additional `New-Folder` cmdlets creating the `NFS` and `iSCSI` datastore folders and the `Finance` and `IT` host folders. These follow the same format:

```
New-Folder -Name "NFS" -Location (Get-Datacenter -Name "Primary" |
Get-Folder -Name "datastore")


New-Folder -Name "iSCSI" -Location (Get-Datacenter -Name "Primary"
| Get-Folder -Name "datastore")


New-Folder -Name "Finance" -Location (Get-Datacenter -Name
"Primary" | Get-Folder -Name "host")


New-Folder -Name "IT" -Location (Get-Datacenter -Name "Primary" |
Get-Folder -Name "host")
```

8. With all of the folder structures now created, you can take a look at moving objects into these locations. You will use multiple cmdlets that begin with `Move-` in order to relocate objects into these folders you have created. Let's begin with the VM folder VMware, and relocate our vCenter server into that folder. Since you know that this is the only folder named VMware in vCenter, you will use the shortcut and just use `Get-Folder` with the name:

```
Move-VM -VM "vCenterSrv" -Location (Get-Folder -Name "VMware")
```

9. The next object you want to relocate is the cluster you created named `BigCluster` using the `Move-Cluster` cmdlet. `Move-Cluster` requires the `-Location` parameter and also a cluster name with the `-Cluster` parameter. You can relocate `BigCluster` into the `IT` host folder:

```
Move-Cluster -Cluster "BigCluster" -Location (Get-Folder -Name
"IT")
```

10. Lastly, you can reorganize our datastores logically using the `Move-Datastore` cmdlet. This cmdlet uses the parameter `-Destination` instead of `-Location`, but accepts the input of a `Folder` object:

```
Move-Datastore -Datastore "NFSDatastore1" -Destination (Get-Folder
-Name "NFS")


Move-Datastore -Datastore "iSCSIDatastore1" -Destination (Get-
Folder -Name "iSCSI")
```

11. Unfortunately, there is no native cmdlet to move `PortGroups` from `Standard vSwitches` into folders in the Networking view, but it can be done in the vSphere Client GUI.

How it works...

The `New-Folder` cmdlet automatically determines what type of folder to create based on the location passed to it. There are five types of folders: datacenter, vm, host, network, and datastore. If you pass the vm folder location, a vm folder will be created. The type of folder determines where the folder is visible in the GUI. In PowerCLI, you can also use these types to scope the results returned from the `Get-Folder` cmdlet by passing the `-Type` parameter.

If you can see, each of the `Move-` cmdlets accepts a name parameter that is specific to the type of object it expects: `Move-VM` uses the `-VM` parameter, `Move-Datastore` uses `-Datastore`, and `Move-Cluster` uses the `-Cluster` parameter. This follows a logical pattern that you can expect even without using `Get-Help` to see instructions.

There's more...

In this recipe, you worked with host, vm, network, and datastore groups within a datacenter. vSphere also allows you to create datacenter folders at the same level as datacenter objects to arrange datacenters logically.

See also

> ▸ The *Deploying new virtual machines from a template* and *Creating basic reports of VM properties using VMware Tools and PowerCLI* recipes in *Chapter 3, Managing Virtual Machines*

Setting permissions on vCenter objects

As a shared computing platform, vSphere has always had a strong roles and permissions model. This allows administrators who control the physical infrastructure and the virtual infrastructure to delegate levels of access to users. vCenter provides nine default roles that you can assign to users on different vSphere objects. By contrast, an ESXi host only has three default roles: Administrator, Read-Only, and No Access.

What is great about the vSphere permission model is that you can take users or groups (both AD, and from vSphere, SSO) and you can assign them a level of access at a cluster, folder, resource pool, datacenter, or at the vCenter root. The same user or group can have different access at different levels, but permissions assigned at a higher level are inherited through objects at lower levels in the hierarchy.

If you have specific needs, vCenter also exposes the ability to create your own roles using individual vSphere privileges. This allows very specific access to be granted for users and tighter security for all of the shared resources. There are hundreds of privileges that can control interaction for vCenter and each one can be configured into a custom role and assigned in vCenter.

In this recipe, you will learn the basic cmdlets used to assign roles and permissions to users and vSphere objects.

Getting ready

To begin, you will need to open a PowerCLI window and connect to a vCenter server.

For the purpose of this recipe, you will take our folder structure and assign groups of users from Active Directory to access these resources using predefined roles. This assumes that you have properly configured your VMware SSO to allow Active Directory authentication.

You will take an Active Directory group called `IT Admins` and delegate access to the entire `Primary` datacenter. You will take the `Finance Developers` group and delegate the operator access to them for the `Finance` folder in vCenter. You will delegate the read-only access to a service account user who is going to be reporting on vCenter using PowerCLI.

How to do it...

In order to set up permissions on vCenter objects, perform the following steps:

1. To begin, you will want to know what roles are available on the system where you can add permissions. To do this, you run the `Get-VIRole` cmdlet and pipe it to a `Select` cmdlet to return just the name and description:

    ```
    Get-VIRole | Select Name, Description
    ```

2. Now that you have a list of roles to work with, you can begin the permission assignment. To do this, you will use the `New-VIPermission` cmdlet. This cmdlet requires an Entity where the permission will be applied, a Principal who represents the user or group and the desired role. For the first cmdlet, you will grant the `Admin` role on the `Primary` datacenter to our `IT Admin` group, which has the principal name `DOMAIN\IT Admin`:

    ```
    New-VIPermission -Entity (Get-Datacenter "Primary") -Principal
    "DOMAIN\IT Admins" -Role Admin
    ```

3. Using the same format for another `New-VIPermission` cmdlet, you can now grant our `Finance Developers` group the operator status as `VirtualMachineUser` on the `Finance` folder. You will use the `Cct Folder` cmdlet to set our entity (or location):

    ```
    New-VIPermission -Entity (Get-Folder "Finance") -Principal
    "DOMAIN\Finance Developers" -Role VirtualMachineUser
    ```

4. Lastly, you want to grant read-only access to the `reports` service account that will be used to script reports and/or perform monitoring on vCenter. You will again use the `New-VIPermission` cmdlet and the `ReadOnly` role granting access to the `Primary` datacenter:

```
New-VIPermission -Entity (Get-Datacenter "Primary") -Principal
"DOMAIN\reports" -Role ReadOnly
```

5. For the next step, suppose you have some users that work at an IT Monitoring office, who need to be able to monitor vSphere and clear alarms on vCenter. There is no predefined role that has those specific permissions. However, with PowerCLI, you can create a new role. To begin, let's use the `Get-VIRole` cmdlet to retrieve the `readonly` role and view the privileges assigned:

```
Get-VIRole -Name readonly | Select Name, PrivilegeList
```

6. With this, you can see that the `readonly` role has three privileges –`Anonymous`, `View`, and `Read`. You can use these as the basis of our custom role. However, you need to give the user rights to clear alarms, so the next step is to find the privilege to do this. In order to find this, you can use the `Get-VIPrivilege` cmdlet to get a list of privileges related to alarms by searching with the help of the `-Name` parameter and a wildcard:

```
Get-VIPrivilege -Name *alarm*
```

7. With the output from the previous cmdlet, you can see a privilege called `Acknowledge alarm` that should allow our `IT Monitoring` group the access they need. The next step is to create our custom role. To do this, you use the `New-VIRole` cmdlet and pass in a name for our new role and the privileges you have found using the `-Privilege` parameter:

```
New-VIRole -Name "IT Monitoring" -Privilege "Anonymous", "View",
"Read", "Acknowledge alarm"
```

8. Last but not least, you have to assign the permission using our new role. To do this, you use the `New-VIPermission` cmdlet again with the `-Entity`, `-Principal`, and `-Role` parameters:

```
New-VIPermission -Entity (Get-Datacenter "Primary") -Principal
"DOMAIN\IT Monitoring Group" -Role "IT Monitoring"
```

How it works...

With the vSphere roles and permissions model, administrators have a very high level of control over what and where users can have access. The default roles can be easily leveraged to assign permissions for common sets of functionality. A role is a defined group of privileges that can be assigned to an individual user or group of users. Privileges are specific rights to perform granular tasks in vSphere. If a default role doesn't have the exact mix of privileges that you need to grant, vSphere is extensible, and a custom role can be created.

The second half of the model is the permissions. Permissions use defined roles along with user account or groups. From a cmdlet standpoint, the user account or groups are known as Principals. A permission consists of a role and a principal, and permissions are defined on particular objects. Permissions are inherited through the hierarchy, which means that if you grant a permission at the datacenter level, then all of the folders, clusters, hosts, virtual machines, networking, and datastores will inherit the permission granted at the datacenter level.

As you illustrated in the recipe, you can take individual folders that contain groups of VMs for a specific group of users and grant permissions for them. In our example, our `Finance Developers` group of users need operational privileges on the Finance folder of VMs. Using the default `VirtualMachineUsers` role, you can grant them access to do operations such as power on and power off the VMs, and use the remote console.

There's more...

With hundreds of privileges packaged with vCenter, it can be daunting to try and create custom roles. Some privileges that might not be obviously required can prevent a custom role from having the desired access. One suggestion is to take an existing or default role and then work from its privilege set to alter it for your uses. This can easily be done by retrieving an existing privilege set and storing them in a variable. Then you can pass this existing list of privileges into your new custom `VIRole`.

For many environments, the default roles can be absolutely sufficient for most administration. The other great advantage of using the default roles is that these change from version to version of vCenter as new privileges can be added.

3
Managing Virtual Machines

In this chapter, you will cover the following recipes:

- ► Deploying the first virtual machine
- ► Cloning a virtual machine to a template
- ► Deploying new virtual machines from a template
- ► Importing a virtual appliance from OVA
- ► Performing a hot add of virtual hardware to an existing virtual machine
- ► Enabling and disabling Fault Tolerance on a virtual machine
- ► Increasing the disk space in a virtual machine
- ► Upgrading the virtual hardware version of a virtual machine
- ► Locating and reloading inaccessible or invalid virtual machines
- ► Setting VMware Tool settings from PowerCLI
- ► Creating basic reports of VM properties using VMware Tools and PowerCLI

Introduction

In the first two chapters of this book, you created a base platform that will allow you to run virtual machines. In this chapter, you begin to work with actual workloads. This chapter will cover provisioning the first VM, deploying virtual machines in bulk, and managing the virtual machines.

There are actually several ways to create a virtual machine. The first is to build it like you would build a normal physical server. For this, you need to install an operating system onto a blank disk from the installation media (CD, DVD, a USB flash drive, or an ISO image). This process is very manual and repetitive for administrators.

vSphere improves this process through the ability to clone a VM once it's built. Cloning makes an identical copy of your VM so that it can be deployed multiple times. Cloning is also important for other tasks such as creating test labs and replicating problems.

In addition to cloning, vSphere can also mark a built VM as a template, signifying that it is prepared with the intent to build other VMs. Templates cannot be powered on or changed. This limits a prepared VM from accidentally being used for another purpose. Cloning and deploying a VM from a template are very similar.

The fourth way to deploy a VM is to deploy a virtual appliance. Based on a very similar technology, deploying a virtual appliance is a workflow that imports a specially formatted file. A virtual appliance often comes in an **Open Virtual Appliance** (**OVA**) or **Open Virtualization Format** (**OVF**). vSphere has a workflow to import OVA and OVF in order to create machines.

However, getting a VM started is only the beginning, because all administrators know that maintaining computers takes much more time and energy than simply deploying them. You will cover several additional recipes to help with the administration and upkeep of your virtual machines once you deploy them.

Deploying the first virtual machine

To begin deploying your first virtual machine from PowerCLI, the first thing you will notice about this recipe is the number of parameters that you are going to need to specify in order to create a new virtual machine. There are a lot of things that go into defining a fully functioning server including basic things such as the number of processors, the amount of RAM, the number and the size of the virtual disks, and a name. Other parameters that are required are going to be specific to the virtual environment, such as defining the host, the VM folder, and the resource pool that the VM is going to reside in.

Getting Started

To begin this recipe, you will need to open a PowerCLI window and connect to a vCenter server.

How to do it...

1. Creating a new VM using PowerCLI uses the `New-VM` cmdlet. The first parameter will have a name, which is the identifier that you will use in other cmdlets to identify this VM. Try to execute the cmdlet and look at the error you receive:

    ```
    New-VM -Name WinVM1
    ```

2. The error tells us that you need to specify a VMHost, ResourcePool, or vApp. The last parameter is actually deprecated, so you really have only two choices. In this example, let's use `Production`, one of the ResourcePools created in *Chapter 2, Configuring vCenter and Computing Clusters*, which is as follows:

    ```
    New-VM -Name WinVM1 -ResourcePool "Production"
    ```

    If you execute the cmdlet now, it will execute successfully and create a VM with 1 vCPU, 0.25 GB of RAM, and a 4 GB hard disk. However, this doesn't meet the needs of running a modern version of Windows. If you have already created a VM, let's delete it now with the `Remove-VM` cmdlet:

    ```
    Get-VM WinVM1 | Remove-VM -DeletePermanently
    ```

3. For the next step, you need to define some additional parameters. First, let's specify the number of CPUs to be assigned and specify the amount of RAM to be dedicated to this new VM. You can do this with the `-NumCPU` and `-MemoryMB` or `-MemoryGB` parameters.

    ```
    New-VM -Name WinVM1 -ResourcePool "Production" -NumCPU 2
    -MemoryGB 4
    ```

4. A Windows VM will need at least 40 GB, or preferably 60 GB, so you will have to add a parameter to specify the disk size. Similar to the RAM allocation, you can specify the size of our disks with `-DiskMB` or `-DiskGB` parameters. You can also specify more than one disk with sizes separated by commas. In this example, you will create a single 60 GB disk that is thick provisioned by default:

    ```
    New-VM -Name WinVM1 -ResourcePool "Production" -NumCPU 2
    -MemoryGB 4 -DiskGB 60
    ```

5. Another important specification is a network so that the VM can communicate. You can specify a port group or multiple port groups that the VM should be connected to using the `-NetworkName` parameter. Multiple port groups should be separated by commas. For this recipe, you will use the Infrastructure Network port group you created in *Chapter 1, Configuring the Basic Settings of an ESXi Host with PowerCLI*, which is as follows:

    ```
    New-VM -Name WinVM1 -ResourcePool "Production" -NumCPU 2

    -MemoryGB 4 -DiskGB 60 -NetworkName "Infrastructure Network"
    ```

6. The cmdlet that we've assembled will certainly create a VM, but something is still missing and that's the operating system. To install an operating system, the easiest thing to do is to attach a bootable ISO image to the VM, but to do this you need to add a virtual CD-ROM drive to the VM. To do this, you will use the `-CD` parameter. If you need a virtual floppy drive, you can also add it with `-Floppy`:

```
New-VM -Name WinVM1 -ResourcePool "Production" -NumCPU 2
-MemoryGB 4 -DiskGB 60 -NetworkName "Infrastructure Network" -CD
```

7. There is one more important thing that needs to be defined and that is the Guest operating system. In vSphere, the OS defined on a VM will allow features that are compatible and disable features that are not compatible. For this recipe, you will set the `GuestID` parameter to `windows7server64Guest`, which is the ID for Windows Server 2008 R2.

```
New-VM -Name WinVM1 -ResourcePool "Production" -NumCPU 2 -MemoryGB
4 -DiskGB 60 -NetworkName "Infrastructure Network" -CD -GuestID
"windows7server64Guest"
```

8. When you execute the cmdlet, you will get a confirmation output that shows the name of the VM, power state, number of CPUs, and RAM, as shown in the following screenshot:

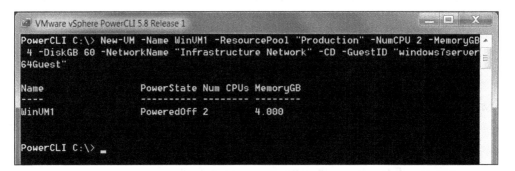

9. Now that the VM is created, you can start the VM using the `Start-VM` cmdlet. The `Start-VM` cmdlet only needs to know which VM to start with. To supply this, you can use the `Get-VM` cmdlet, which is as follows:

```
Get-VM WinVM1 | Start-VM
```

You could have actually started the VM by piping the `Start-VM` cmdlet with no additional parameters to our `New-VM` cmdlet. The following command line is an example:

```
New-VM -Name WinVM1 -ResourcePool "Production" -NumCPU 2 -MemoryGB
4 -DiskGB 60 -NetworkName "Infrastructure Network" -CD -GuestID
"windows7server64Guest" | Start-VM
```

10. The last step is to mount an ISO file to the CD drive so that the operating system can begin the installation. In this example, you will use an ISO file called `Windows2012Server.iso` that is stored on the `iSCSIDatastore1` datastore. To mount the CD, you will use the `New-CDDrive` cmdlet:

    ```
    New-CDDrive -VM WinVM1 -IsoPath "[iSCSIDatastore1]
    Windows2012Server.iso"
    ```

11. At this point, the VM should recognize and begin booting from the CD drive, but you might need to interact with the console, so you should load a remote console session. You can do this with the `Open-VMConsoleWindow` cmdlet:

    ```
    Get-VM WinVM1 | Open-VMConsoleWindow
    ```

How it works...

The `New-VM` cmdlet taps into the normal GUI workflow to create a VM, but where the GUI presents a step-by-step group of choices to define the options, the PowerCLI `New-VM` cmdlet requires that all of the decisions be made upfront. The `New-VM` cmdlet can work with fewer parameters, but if you have specific needs, there are parameters that let you customize the VM to your needs.

Additionally, you looked at the `New-CDDrive` cmdlet that allows you to take an ISO image that is stored on a datastore and mount it to the VM you created. This maps the file to the virtual CD drive that you defined in the `New-VM` cmdlet. This allows you to boot and install an operating system into the VM. Finally, you used the `Open-VMConsoleWindow` to allow you to interact with the remote console and complete the installation.

I will certainly concede that using the GUI to build your first VM is probably the easiest way to build a single VM. As you will see in the subsequent recipes, understanding how to create virtual machines using PowerCLI is essential to other processes that you will perform more frequently in PowerCLI.

There's more...

As you complete this first recipe, you might wonder why you would ever choose to use PowerCLI to create a blank virtual machine. Cloning a VM from a template or deploying a virtual appliance seems much more usable since these two options will include an operating system. The reason the first option is included is to cover use cases beyond static operating system deployments. There are a number of use cases where the guest operating system you wish to run might be deployed from a PXE boot directly into a virtual machine.

Virtual Desktop Infrastructure (**VDI**) and even farms of application servers might boot from dynamically assigned boot images and boot over the network. These types of PXE boot environments allow administrators to quickly increase or decrease the number of servers that deliver an application. In the VDI use case, you might need to redeploy a patched image and instead of having to patch 100 virtual desktops, you will only need to patch a single image and then reboot all 100 virtual desktops to update them from their golden master delivered over a network boot.

Once again, when it comes to doing work in bulk, you can easily seed an array of integers to a `ForEach` loop and quickly deploy 100 or 1,000 virtual machines. All uniform settings are ready to boot from the network and you can accomplish deploying all of these virtual machines with less than 10 lines of PowerCLI code. It would take 15 to 20 clicks per VM to deploy these in the GUI.

See also

> ▸ VMware SDK Documentation: **Enum – VirtualMachineGuestOsIdentifier** is available at `http://pubs.vmware.com/vsphere-55/topic/com.vmware.wssdk.apiref.doc/vim.vm.GuestOsDescriptor.GuestOsIdentifier.html`

Cloning a virtual machine to a template

Cloning is one of the golden features of virtualization. It is a feature that once you have used it, you will never want to go back to manually deploying servers. It really eases the problems of deploying consistent and compliant virtual machines by allowing an administrator to create a good, compliant image and then make copies of it for all future deployments.

While you can clone regular virtual machines, and there might be times when you need to do this, most of the time you will be cloning new virtual machines from a template. In this recipe, you will take a prepared VM with its operating system installed and basic configuration set, and explore how to clone it and how to convert it to a template.

Surprisingly, there is no `Clone-VM` cmdlet. The clone functionality is actually rolled into the `New-VM` cmdlet, which makes sense because essentially, a cloned VM is a new VM.

Getting Started

To begin this recipe, you will need to open a PowerCLI window and connect to a vCenter server. In addition, you will need to have at least one VM created in the installation.

How to do it...

1. To begin, you will first retrieve an existing VM using the `Get-VM` cmdlet. This VM is going to be used as our template, so you need to get it in order to convert it:

   Get-VM WinVM1

2. Next, you want to make a copy of the VM that will become a template. You will use the `New-VM` cmdlet again to create a clone of the VM. You will specify a name, `WinTemplate`, for the clone. Since the VM won't be running, you will also want to try and conserve the disk space, so you will want to create a thin provisioned disk with the `-DiskStorageFormat` parameter. Finally, you must specify either `-VMHost` or `-ResourcePool`, and in this example you will reuse our `Production` ResourcePool.

   Get-VM WinVM1 | New-VM -Name "WinTemplate" -DiskStorageFormat Thin -ResourcePool Production

3. After you deploy the clone, the next step is to convert our new VM into a template. For this, there is a special parameter with the `Set-VM` cmdlet named `-ToTemplate` that simply converts the specified VM to a template:

   Set-VM -VM "WinTemplate" -ToTemplate -Confirm:$false

How it works...

When using the `New-VM` cmdlet to clone, you need to pass in the name of a VM to clone from or you can pass it through the pipeline. The `-VM` parameter is used when you specify it in the command line. In this recipe, you retrieved a VM object using the `Get-VM` cmdlet and passed it through the pipeline. Because each object passed through the pipeline had a specified type, PowerCLI knew that the input was to be used for the `-VM` parameter. Since this parameter expects a VM object or a string, it can be used to search for a VM object. You can easily rewrite the cmdlet using the `-VM` parameter, as shown in the following command line:

New-VM -Name "WinTemplate" -DiskStorageFormat Thin -ResourcePool Production

The conversion of the VM that you cloned into a template is very simple. It takes a simple `Set-VM` cmdlet with the special `-ToTemplate` parameter created for this specific task. However, since this cmdlet uses the `Set-VM` cmdlet, there are actually many additional things that can be done at the same time as the conversion. You can change parameters, such as the number of CPUs and amount of RAM, using the same parameters that you used in the previous recipe when you created new virtual machines.

There's more...

There is another free resource that VMware provides and that is the online PowerCLI Documentation at the VMware website. This reference provides the same information that you can get from the `Get-Help` cmdlet in an easy and searchable online format. One of the things useful at the online documentation is the table of all the parameters that includes a description, whether it accepts an input from the pipeline and whether it is a required parameter.

See also

▶ VMware, **vSphere PowerCLI Documentation** is available at `https://www.vmware.com/support/developer/PowerCLI/`

Deploying new virtual machines from a template

Deploying a new virtual machine from a template is surprisingly easy. This is a task that you will perform often. Although there are some template specific cmdlets, these have to do with making changes to templates after they are converted. To deploy a VM, you come back to `New-VM` cmdlet.

Getting Started

To get started, you should open a new PowerCLI window and connect to the vCenter server where you defined our template VM.

How to do it...

1. To begin this recipe, you will need to assemble a `New-VM` cmdlet. The first step is to specify the template that is to be cloned from using the `-Template` cmdlet. As of vSphere 5.5, the `-Template` parameter can accept pipeline input, but this is deprecated, so it is better to specify the template by a parameter:

```
New-VM -Template "WinTemplate"
```

2. The next step is to add the name for the VM and the host or the ResourcePool that the VM is going to deploy into:

```
New-VM -Template "WinTemplate" -Name "NewWinVM" -ResourcePool
"Production"
```

3. Since the template is thin provisioned, you might also want to convert this back to a thick provisioned disk. This is sometimes a recommendation for storage that already has thin provisioning built in at the array level. Again, this is done with the `-DiskStorageFormat` parameter. For thick disks, our choices are `EagerZeroThick` or `Thick`, which is the Lazy Zeroed Thick option:

```
New-VM -Template "WinTemplate" -Name "NewWinVM" -ResourcePool
"Production" -DiskStorageFormat Thick
```

4. Not quite complete, yet, you can take a shortcut by specifying a folder location for the new VM. You will assume that `NewWinVM` is an application server and needs to be placed in `App Servers`:

```
New-VM -Template "WinTemplate" -Name "NewWinVM" -ResourcePool
"Production" -DiskStorageFormat Thick -Location "App Servers"
```

How it works...

For the third recipe, we've turned to the `New-VM` cmdlet. It's important to note, however, that you have used `New-VM` in three different ways and each way has a different set of parameters that can be used. Take a look at the following output from `Get-Help New-VM`:

You will see that `New-VM` has four groupings of parameters that work together, but not all of the cmdlets are accepted in all parameters. This particular use case with the `-Template` parameter doesn't accept inputs such as `-NumCPU` and `-MemoryGB` and will throw an error if specified.

However, why? These are valid parameters for the New-VM cmdlet, correct? Yes and no. When a cmdlet plays many roles, such as New-VM, certain things are not possible. When you provision a VM from a template in the GUI, that workflow doesn't include the ability to change the hardware during the deployment.

 Changing the VM hardware has existed as an experimental feature in vSphere for a few versions, but it isn't a part of the same deployment workflow.

Since this is not an available option, the parameters used for deploying a template do not accept the parameters that would alter the VM hardware profile. You will encounter a number of these multi-use cmdlets in PowerCLI which is why you are referred to Get-Help and online documentation. These resources will clear syntax questions such as these.

There's more...

Deploying a virtual machine from a template is a fantastic feature of vSphere. However, it doesn't fully address the problem of deploying virtual machines because the guest operating system will need further customization to truly become a different virtual machine. Virtual machines running Windows operating systems will require the Sysprep process to be run, which generates a new **System Identifier** (**SID**), making it a new server. Virtual machines running Linux will need to customize the network settings and reset these since the **Media Access Control** (**MAC**) address of the virtual machine has changed and it will need to be reconfigured for network connectivity.

To handle some of these use cases, vSphere packages a customizer feature that allows an administrator to define the basic settings to be applied against a guest operating system. These settings can include the network settings, the name of the machine, licensing information, and even the ability to change the guest's virtual hardware during deployment.

You will see that two of the parameter sets include a -OSCustomizationSpec parameter. This parameter allows you to pass an OS customization specification to the template to be executed after the VM deploys. The OS customization specification is a set of parameters, including settings to change the SID, the administrator password, the domain information to join the system to the domain, network settings, product key, time zone, and other settings. The OSCustomizationSpec object contains all of this information, so you can create an object with the Set-OSCustomizationSpec cmdlet and pass this in as a variable to the New-VM cmdlet.

Importing a virtual appliance from OVA

Not all virtual machines have to be built from scratch or built from a template that you've created in-house. Many virtual machines are distributed in appliance form using the OVA and OVF formats. OVA is a single file that contains all of the details and virtual disk information for a virtual machine.

OVF is a set of files that contains specifications and the data disks for a virtual machine. These two formats allow vendors to create pre-defined copies of their application and easily distribute them for use.

Importing a virtual appliance from PowerCLI is actually a pretty simple task to accomplish, but it does come with a bit of risk. Not all virtual appliances are created in a way such that importing them from PowerCLI will work. For instance, the **vCenter Server Appliance** (**VCSA**) is not a great candidate for importing in PowerCLI because it requires a lot of additional configuration questions to be answered during the import wizard. Without these settings being defined, the initial boot and configuration of the VM will fail.

However, there are many other applications that are easily imported from PowerCLI. Load balancers, web servers, and other applications are distributed and they require no additional customization. These simple appliances boot the first time, obtain an address from DHCP, and then allow you to perform the configuration. For these, PowerCLI offers a significant time saving when deploying multiple copies of a virtual appliance. However, you should test the virtual appliance first in the GUI in order to know if additional custom properties are required for a successful deployment, and if they are not, you can proceed with PowerCLI.

When you are researching solutions, the **VMware Virtual Appliances Marketplace**, page located at `https://solutionexchange.vmware.com/store/category_groups/virtual-appliances`, is a great resource to find virtual appliances.

Getting Started

To begin this recipe, you will need to open a PowerCLI window and connect to a vCenter server. You will also need to find a virtual appliance and download it so that you can use it for this recipe. If you need a virtual appliance, but you aren't sure what to download, blogger Mike Laverick has several options available for download at his website, `http://www.mikelaverick.com/download/`. Another possibility is SmartOS from Joyent that is available at `http://www.smartos.org`.

For this recipe, you will use a virtual appliance called `SliTaz 4.0`, which is a small OVF and OVA downloaded from Laverick's website.

How to do it...

1. To begin, note the location where you have saved your files for the virtual appliance. If the appliance download is zipped, unzip it and save it in an easy to access location. For this recipe, my virtual appliance is located at `C:\va` on my local machine.

2. Next, you will use the `Import-VApp` cmdlet to import the appliance files. The first and the most important parameter is the `-Source` parameter that points to the OVF or the OVA file. You must also specify a host using `-VMHost`:

    ```
    Import-vApp -Source C:\va\SliTaz4.0\SliTaz4.0.ovf -Name
    "SliTaz4.0" -VMHost esxhost1.domain.local
    ```

3. You will see a progress bar as the import completes, followed by the output with confirmation of the `Name`, `PowerState`, Number of CPUs, and the amount of memory assigned to your imported virtual appliance:

```
VMware vSphere PowerCLI 5.8 Release 1                          _ □ X

PowerCLI C:\va\SliTaz4.0> Import-vApp -Source C:\va\SliTaz4.0\SliTaz4.0.ovf -Nam
e "SliTaz4.0" -VMHost esxhost1.domain.local

Name                    PowerState Num CPUs MemoryGB
----                    ---------- -------- --------
SliTaz4.0               PoweredOff 1        0.125


PowerCLI C:\va\SliTaz4.0> _
```

4. Once the deployment finishes, you can power on the new VM. To do so, just use the `Start-VM` cmdlet with the `-VM` parameter to specify our new `SliTaz4.0` VM:

    ```
    Start-VM -VM "SliTaz4.0"
    ```

How it works...

When you execute the `Import-vApp` cmdlet in PowerCLI, the distributed virtual appliance files are read in using the `-Source` cmdlet and validated against the hash provided as a part of the package. Once the validation occurs, the virtual disk is uploaded to a datastore, either specified by a parameter or chosen by vSphere if not specified. In our recipe, you did not specify a particular datastore for the VM to be placed, so vSphere chose the datastore location automatically on the specified host. The `-VMHost` parameter is mandatory and you must deploy to a host specifically. You can also specify a `-Location` value that can be any sort of VIContainer, such as ResourcePools, Clusters, or Folders are all possible, but it is not a replacement for `-VMHost`.

Importing a virtual appliance or a vApp is simple from a PowerCLI perspective. What goes on behind the scenes is a bit more than it might appear. Each OVA file or OVF bundle also bundles a hash that is used to verify the validity of the image before it is imported and deployed. If the hash does not match the data files provided, an error is presented and the import will fail. This is intended to stop corruption and to avoid malicious changes from being made to the virtual appliance once it's created.

Although this section focuses on importing a virtual appliance, the same cmdlets can be used to deploy a vApp. A vApp is nothing more than several bundled virtual appliances that work together to form an application. One virtual appliance can be a database server or a collection server while the other is a management server. Both are needed for a fully functioning application, so that they are bundled together using the vApp construct in vSphere.

A great example of a vApp is **vCenter Operations Manager** (**vCOPS**). vCOPS is a multiserver application that bundles two virtual appliances: a UI virtual machine and an analytics virtual machine. The import process happens as a single vApp but in the background it deploys both the virtual machines.

There's more...

As you can imagine, deploying virtual appliances and vApps in bulk is particularly handy. With an easy `For` loop, you can deploy many of these for lab or classroom environments to quickly get workloads up and running as you need. Beyond testing environments, deploying many virtual appliances is handy in production environments when deploying clusters of virtual network appliances, such as load balancers.

See also

- ► Mike Laverick's blog contains OVF/OVA downloads at `http://www.mikelaverick.com/download/`
- ► SmartOS from Joyent is available at `http://www.smartos.org`

Performing a hot add of virtual hardware to an existing virtual machine

One of the greatest benefits of virtualization is the ability to give a virtual machine a hardware upgrade without buying new hardware. Since each virtual machine only uses a portion of the host's available resources, you can reconfigure the virtual machine and add additional hardware. You can add a USB controller, a CD drive, additional hard disks, SCSI controllers, vCPUs, and RAM to the virtual machines that have already been configured.

In the past, adding CPU and RAM to a virtual machine was a task that had to be done while the VM was offline, but increasingly, guest operating systems are establishing support for hot add, which means that administrators can increase the RAM and number of CPUs of a virtual machine while the guest is running.

The most recent version of Windows and Linux support hot add vCPU and RAM. VMware publishes a Compatibility Guide page for Guest OS, which outlines features available for each version of the supported guest OS, including whether it can support vCPU and memory hot add. The link for the compatibility guide is at the end of this recipe. From a GUI standpoint, you go to the same **Edit Settings...** dialog box and when you select RAM, the settings are no longer greyed out. After you change and save the settings, the virtual machine sees and recognizes the additional memory.

In this section, you will learn about adding hardware to a virtual machine. You will also look at which functions will work online and which will require the VM to be offline in order to make the change.

Getting Started

In this recipe, you will run through a number of common hardware reconfiguration that needs to occur in virtual machines and can easily be handled through PowerCLI. The first thing you will cover are some online reconfigurations: adding vCPU and memory to a supported Guest OS, adding disk space to an existing disk, and adding a brand new virtual disk to an existing VM. Next, you will cover some reconfiguration that must be done offline such as adding an additional SCSI controller and a different set of virtual disks needed to set up a Windows cluster running in VMware. To begin, you will need to open a PowerCLI window and connect to a vCenter instance.

How to do it...

1. To begin, you have to figure out which virtual machine needs to be reconfigured. You will do this with the Get-VM cmdlet:

   ```
   Get-VM Win*
   ```

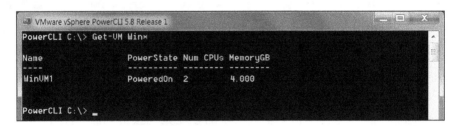

2. You will see the output that indicates WinVM1 is powered on. You will use the Set-VM cmdlet to make changes to the running VM. You will also see that WinVM1 is running with two CPUs and 4 GB of RAM. However, 4 GB is not sufficient for the application you are running, so you need to increase this amount to 12 GB. To do this, you will use the -VM parameter to specify the VM name and the -MemoryGB to reallocate more RAM:

   ```
   Set-VM -VM WinVM1 -MemoryGB 12
   ```

When you execute the cmdlet, you get an error because hot add is not enabled, which is shown as follows:

```
The operation for the entity "WinVM1" failed with the following
message: "Memory hot plug is not supported for this virtual
machine."
```

3. The reconfiguration failed in this case because you did not define the running VM with the ability to hot add RAM or CPUs. You will need to shut down the VM and reconfigure it. However, there is no cmdlet specifically to do this. You will need to use the `Get-View` cmdlet. The first step is to shut down the VM with `Stop-VM` cmdlet and then retrieve a View of the VM with `Get-View`.

```
Stop-VM -VM WinVM1
$vmview = Get-VM WinVM1 | Get-View
```

4. The next step is to create a new configuration specification object and store the settings you want to change with their needed settings. To do this, you will first define a new object with the type, `VMware.VIM.VirtualMachineConfigSpec`, which is the object type for a virtual machine's configuration:

```
$ConfigSpec = New-Object VMware.VIM.VirtualMachineConfigSpec
```

5. The next step is to set the two options that control the hot add capability. To do this, you need one more object to be defined: `VMware.VIM.optionvalue`:

```
$options = New-Object VMware.VIM.optionvalue
```

6. The next step is to define key and value pairs for the options you want to change. Our keys for hot add are `mem.hotadd` and `vcpu.hotadd`. You will set both of these to a value, `true`:

```
$options.Key = "mem.hotadd"
$options.Value = "true"
```

7. The next step is to add this pair to the configuration specification. You can do this by using the `+=` assignment operator:

```
$ConfigSpec.extraconfig += $options
```

8. The last step is to commit these changes to the VM. For this, you will use the VMView object's built-in `Reconfig()` function. You can call the VMView object using the `$vmview` variable you defined earlier. You pass the `$ConfigSpec` variable you created into this view to redefine these options:

```
$vmview.ReconfigVM($ConfigSpec)
```

9. Next, you repeat the last three steps with a new key and value pair for vCPU hot add:

```
$options.Key = "vcpu.hotadd"
$options.Value = "true"
$ConfigSpec.extraconfig += $options
$vmview.ReconfigVM($ConfigSpec)
```

10. At this point, you can restart the VM with the `Start-VM` cmdlet:

    ```
    Start-VM -VM WinVM1
    ```

11. Once it has a chance to get back online, you can attempt to hot add RAM to the VM:

    ```
    Set-VM -VM WinVM1 -MemoryGB 12
    ```

    This could have been executed while the VM was offline without the need to reconfigure the VM for hot add vCPU and memory, but if you need to add RAM or vCPU in the future, the VM is ready to do this while it's online.

12. The same application needs an additional data disk with 20 GB of disk space. You will add this as a new disk on the VM:

    ```
    Get-VM WinVM1 | New-HardDisk -CapacityGB 20
    ```

13. Once the disk is added to the VM, the operating system will need to format and prepare the disk for use.

How it works...

In this recipe, you attempted to perform a hot add or an online add of RAM into a VM. The VM was not configured to enable the hot add feature, even though the operating system supported the feature. When you created WinVM1 earlier in this chapter, you created it with the GuestID for Windows Server 2008 R2, which supported the hot add vCPU and memory. However, the feature does not get enabled by default. What's more, there isn't a cmdlet to enable this feature in a VM either. Within the `Set-VM` cmdlet, there are no parameters to enable hot add. So, you have to turn to lower-level SDK features to achieve this.

All of the high-level cmdlets in PowerCLI leverage the same underlying web SDK and make calls against it to perform the operations that you have executed. However, in this example, it is up to us to manually retrieve and create objects of the `VirtualMachineConfigSpec` and the `optionvalue` types. Once you create these with the `New-Object` cmdlet, you are able to populate data into these that you can use in conjunction with a `VMView` object retrieved with the `Get-View` cmdlet.

There's more...

As PowerCLI has matured, there have been many cmdlets that have been added to the toolset. In the past, it was necessary to work with Views and SDK-based objects to achieve a number of setting changes. There is tremendous power working with Views too. It is impossible to cover all of the possibilities within PowerCLI, simply because it is an extensible and flexible toolkit where administrators and developers can build their own functionalities easily.

If ever you hit a roadblock trying to achieve something that you need to do with PowerCLI, there is a good chance that someone has created a script leveraging the lower-level SDK Views and functions in PowerCLI. It is always a good bet to start with a web search to try and discover the resources others are making available for free. While it is not nearly as straight forward as working with native cmdlets, the potential is there if you need to script a setting change.

See also

▶ VMware Guest OS Compatibility Guide is available at `http://www.vmware.com/resources/compatibility/pdf/VMware_GOS_Compatibility_Guide.pdf`

Enabling and disabling Fault Tolerance on a virtual machine

In addition to VMware **High Availability** (**HA**) clustering, VMware also provides the ability to enable **Fault Tolerance** (**FT**) to protect a running virtual machine. FT creates a secondary virtual machine on a second host in the cluster and executes all of the same instructions on both the VMs. In the event of a host failure on the server hosting the primary VM, the secondary VM will assume the role of the primary with no downtime. There are no breaks in the network connectivity or application uptime.

In vSphere 5.5, FT has a significant list of requirements, including at least two ESXi hosts with FT compatible processors, shared datastores, and networking hosting the VM. In addition, your ESXi hosts must have a logging network connection between the hosts for Fault Tolerance where the process instructions are transmitted for execution on the secondary VM. Most restrictive, however, is vSphere 5.5 which only supports FT on a virtual machine with only one vCPU.

If you meet these requirements, however, enabling FT on a virtual machine is very simple from PowerCLI. There is not a native cmdlet to handle this, but the `ExtensionData` property in a VM object has a method to enable and to disable the protection feature.

Getting Started

For this recipe, you will need a PowerCLI window with a connection to a vCenter Server. You will also require at least two ESXi hosts connected to a shared datastore hosting WinVM1 or another VM whose name can be substituted in the recipe. If the ESXi hosts were built by the commands created in *Chapter 1, Configuring the Basic Settings of an ESXi Host with PowerCLI*, the Fault Tolerance logging network should be created and active between the ESXi hosts.

How to do it...

1. The first step is to retrieve a VM object using the `Get-VM` cmdlet. For an easy reference, assign this object to the `$vm` variable:

   ```
   $vm = Get-VM "WinVM1"
   ```

2. Now that you have the object, you should check to see whether Fault Tolerance is compatible for this VM. To do this, there is a method in `ExtensionData`. Perform a `Get-Method` cmdlet, or `GM` for short, to look inside the `ExtensionData` property of the object:

   ```
   $vm.ExtensionData | GM
   ```

3. If you look through the list of methods in properties in `ExtensionData`, you will find a `QueryFaultToleranceCompatibility` method. This can be used to check for problems that would prevent Fault Tolerance to be enabled. If there are no faults listed, you can continue with the next step. If there are faults listed, these will indicate problems you should resolve before trying to enable FT on the VM:

   ```
   $vm.ExtensionData.QueryFaultToleranceCompatibility()
   ```

4. You can look through the list of methods in `ExtensionData` again with `Get-Method` and to locate the two methods you will use to enable and disable FT. The method used to enable FT is named `CreateSecondaryVM`. The method to disable FT is named `TurnOffFaultToleranceForVM`. With WinVM1 powered off, run the `CreateSecondaryVM` command. You will pass in `$null` for the value of the method that will allow DRS to place the secondary VM onto an ESXi host. Alternatively, you can specify the ID of a specific host where the secondary VM should be placed in the method:

   ```
   $vm.ExtensionData.CreateSecondaryVM($null)
   ```

5. At this point, you can power on the VM and you will see both the primary and secondary VM being powered on in vCenter:

   ```
   Start-VM -VM "WinVM1"
   ```

6. Even while the VM is running, you can turn off Fault Tolerance with the `TurnOffFaultToleranceForVM` method mentioned in Step 4. This is disabled and then removes the secondary VM:

   ```
   $vm.ExtensionData.TurnOffFaultToleranceForVM()
   ```

How it works...

The primary methods for enabling and disabling Fault Tolerance in vSphere are `CreateSecondaryVM` and `TurnOffFaultToleranceForVM`, and these methods are located in the VM object within PowerCLI. `CreateSecondaryVM` will create the secondary VM and then replicate its power state.

If powered on, it will take a snapshot of the VM's memory, start it at a known point, and then begin replaying the instructions from the snapshot forward.

In addition to `CreateSecondaryVM`, you might have seen additional methods called `EnableSecondaryVM` and `DisableSecondaryVM`. Although Enable and Create might sound like similar operations, they have different purposes. `DisableSecondaryVM` is used to disable FT on the next power on of the VM, and `EnableSecondaryVM` enables FT on the next power on of the VM and the method is only used to enable a VM that has been disabled with the `DisableSecondaryVM` method.

When you explore the `ExtensionData` property with `Get-Method`, you will also see almost identically named methods `CreateSecondaryVM_task` and `TurnOffFaultToleranceForVM_task`. These methods generate tasks, but they do not return useful feedback from the command. The `_task` methods only return the identifying value of the task generated on vCenter Server, however, the _task methods perform the same work as their counterparts.

In the recipe, you created the secondary VM on a powered off WinVM1. This was done for simplicity in the recipe, however, sometimes enabling FT on a powered on VM will result in an error "Replay is unavailable for the current configuration." The workaround for that error is to shut down the VM to be protected, so this was suggested prior to running the `CreateSecondaryVM` method for this reason.

See also

▶ VMware vSphere 5.5 Availability Guide is available at `http://pubs.vmware.com/vsphere-55/topic/com.vmware.ICbase/PDF/vsphere-esxi-vcenter-server-55-availability-guide.pdf`

Increasing the disk space in a virtual machine

In the previous recipe, you covered configuring a VM for hot adding vCPU and RAM on an existing virtual machine. In this recipe, you will dive a bit deeper into space allocation for virtual machines. With applications and data demands growing, it's a common use case to need to increase disk space on a virtual server, but it's not just a VMware vSphere setting. In all the cases, you will need to do things within the guest operating system to recognize and utilize the additional space provided.

In terms of virtual hardware, virtual hard disks have always been hardware that can be added while the system was running, just as you did in the previous recipe. However, in addition to adding a hard disk, the sizes of the disks can also be increased while the system is online and this allows you to allocate more disk space without the need to stop an application.

Getting Started

For this recipe, you will need a PowerCLI window with a connection to a vCenter server where WinVM1 was created.

How to do it...

1. The first step is to locate the hard disk that you want to grow. The best way to do this is to locate the VM with a `Get-VM` and pipe that into a `Get-HardDisk` cmdlet:

    ```
    Get-VM "WinVM1" | Get-HardDisk
    ```

2. You will see that more than one disk is returned. Unless you want to grow both to the same size, you will need to select one of the two using a `Where` clause on our previous cmdlet:

    ```
    Get-VM "WinVM1" | Get-HardDisk | Where {$_.CapacityGB -eq 60}
    ```

3. Now that the result is scoped down to the one disk that you want to work with, you can use the `Set-HardDisk` cmdlet to change the size on this disk. To do this, you simply pass the `-CapacityGB` in as a parameter with a new value. When you execute the command, it will ask you for a confirmation, or you can add the `-Confirm:$false` parameter to suppress this:

    ```
    Get-VM "WinVM1" | Get-HardDisk | Where {$_.CapacityGB -eq 60} |
    Set-HardDisk -CapacityGB 80 -Confirm:$false
    ```

4. You can confirm the disk size change with the `Get-VM` cmdlet:

   ```
   Get-VM "WinVM1" | Get-HardDisk
   ```

5. The next step is to go into the operating system and let the filesystem grow or rescan the filesystem in order to recognize the additional disk space.

How it works...

The `Set-HardDisk` cmdlets change the characteristics of the VMDK file under the virtual machine. Each virtual disk is actually made up of two files, one with the metadata that describes the virtual disk and one with the actual blocks of data inside it. For thin provisioned disks, the metadata is updated but the true data file doesn't change since the blocks are allocated only when used. For thick provisioned, both Lazy Zero and Eager Zero disks, the sizing of the data file also changes.

Once the virtual disk files are updated, the virtual machine sees that the disk size changes within the guest operating system. It is up to the operating system to then make use of the additional space by rescanning the disk subsystem and by changing the partition sizing using native OS utilities.

Upgrading the virtual hardware version of a virtual machine

From release to release, VMware has enhanced the functionality of virtual machines and each new set of capabilities is represented by a virtual hardware version number. Each virtual machine has an assigned virtual hardware number and this represents the feature set that the virtual machine can make use of and defines which version of host is required to run the virtual machine. In short, it's basically the file format of the virtual machine, since every virtual machine is made up of a definition file, virtual hard disks, and a few other special purpose files. In order for a virtual machine to make use of the new functionality, its virtual hardware must be upgraded.

Migrating virtual machines from one virtual hardware version to another is another common task during the life cycle of a virtual machine. You might need to add a type of hardware that only works with a virtual hardware version higher than you are running, or you might simply need to stay updated for other technical reasons. Regardless of the reasons, you can bulk change virtual hardware versions from PowerCLI which is a huge time saver for large virtual environments.

Getting Started

To begin, you will need a PowerCLI with a connection to a vCenter Server. You will clone the virtual appliance running an earlier version of the virtual hardware that you will use to upgrade to a later version of the virtual hardware in this recipe.

How to do it...

1. To begin, you will need to do a couple of steps to prepare this recipe. The first step is to import a virtual appliance and specify an older version of virtual hardware so that you have something to upgrade. To do this, you will reuse the New-VM cmdlet from the *Cloning a virtual machine to a template* recipe, which is as follows:

   ```
   New-VM -VM SliTaz4.0 -Name SliTazTest -ResourcePool Production
   ```

2. Next, confirm that the virtual machine you've cloned is running an earlier version of virtual hardware using Get-VM with a Select cmdlet:

   ```
   Get-VM SliTazTest | Select Name, Version
   ```

3. VMware recommends that you take a snapshot of the VM before upgrading the virtual hardware so that you can revert in the event of a problem after the upgrade. To do this, you will use the New-Snapshot cmdlet:

   ```
   Get-VM SliTazTest | New-Snapshot -Name "Before Virtual Hardware
   Upgrade"
   ```

4. The next step is to perform the virtual hardware upgrade:

   ```
   Set-VM -VM SliTazTest -Version v8
   ```

5. Next, you will start the virtual machine again with Start-VM:

   ```
   Start-VM -VM SliTazTest

   Open-VMConsoleWindow SliTazTest
   ```

6. The last step is to verify that the VM is booted and is operational after the upgrade.

How it works...

The actual process of changing the virtual machine version is quite simple. It's a native PowerCLI cmdlet and the syntax is easy to understand. Behind the scenes, the VMX file undergoes a small change, but changing the version doesn't do a lot inside the virtual machine. Any changes will be detected on the first boot of the operating system.

For the best luck with virtual hardware upgrade, VMware recommends that you first upgrade VMware Tools inside the virtual machine. This ensures that you will get all of the latest drivers as a part of the VMware Tools package. The drivers will be required if you add newly available virtual hardware that the virtual hardware version unlocks.

Upgrading virtual hardware from Version 4 or higher is also reversible, but you must make a snapshot of the virtual machine first, as you did in this recipe. In the event that something happened and virtual machine does not function after the upgrade, you can revert to the snapshot quickly to recover. Upgrading from Version 3 is irreversible, even with a snapshot.

So, why are virtual hardware versions or virtual machine versions so important? These versions dictate what version of ESXi you must be running at a host level to run the virtual machine. Version 4 is the most compatible version that can run on ESX or ESXi 3.5 or higher. Version 7 requires ESX or ESXi 4 or higher to run. Version 8 requires ESXi 5 or higher and Version 9 requires ESXi 5.5 to run. Virtual appliances are likely going to come with an older version of virtual hardware since it takes a period of time before some IT shops upgrade their production vSphere version. It's not uncommon to receive virtual appliances running Version 7, such as the SliTaz 4.0 virtual appliance in this recipe.

However, it's more than just which version of ESXi is required. Higher versions unlock new hardware capabilities, increase RAM allocations, support new NIC types, and other features that earlier versions of ESXi didn't support. Paravirtual SCSI controllers, which provide faster disk IO in virtual machines, require version 7 or higher. E1000e network cards in a virtual machine require Version 8 or higher. Version 9 unlocks up to 512 MB of video memory in a virtual machine. Version 10 adds virtual SATA controllers to virtual machines. These are just a few examples, but VMware has an extensive list of differences in vSphere Documentation Center.

See also

▶ VMware vSphere 5.5 Documentation Center is available at `http://pubs.vmware.com/vsphere-55/index.jsp`

Locating and reloading inaccessible or invalid virtual machines

Since virtual machines are nothing more than a group of files that define a computer, sometimes an ESXi host or vCenter might lose communication with a filesystem or datastore where the virtual machine files are stored. In that event, the VM can be labeled as inaccessible or invalid inside vCenter. One way to solve this is to unregister and reregister the virtual machine's VMX file to vCenter or the ESXi host.

When you have many of these inaccessible or invalid VMs, like after a storage outage, it can become a tedious process to cleanup. PowerCLI can offer an easy and quick way to provide a remedy for this situation. In this recipe, we'll explore how to locate and identify inaccessible or invalid virtual machines and how to automatically reregister them to fix the problem.

This recipe is available online and is a good use case that will help you understand the use of View objects in PowerCLI. Many of the View objects in PowerCLI possess a lot of methods or functions, many more than the objects used by native cmdlets.

Getting Started

To begin this recipe, you will need to open a PowerCLI window and connect to a vCenter Server.

How to do it...

1. To begin with this recipe, you will need to get a VMView object. You can do this using the Get-View cmdlet in PowerCLI. This time, you will obtain a VMView object that contains all of the virtual machines connected to this vCenter server:

   ```
   $vmview = Get-View -ViewType VirtualMachine
   ```

2. Now that you have the view, let's explore it a bit more. You will pipe our View object into Get-Member to see which methods and properties are a part of this object:

   ```
   $vmview | Get-Member
   ```

3. Next, let's obtain a VM object from a native PowerCLI cmdlet using the Get-VM cmdlet:

   ```
   $vmobject = Get-VM
   $vmobject | Get-Member
   ```

4. If you measure the objects, you will see that there are 36 properties in $vmview and 38 properties in $vmobject, as seen in the following screenshot:

```
VMware vSphere PowerCLI 5.8 Release 1                                _  □  x
PowerCLI C:\> $vmview | Get-Member | Where {$_.MemberType -eq "Property"} | Meas
ure


Count     : 36
Average   :
Sum       :
Maximum   :
Minimum   :
Property  :


PowerCLI C:\> $vmobject | Get-Member | Where {$_.MemberType -eq "Property"} | Me
asure


Count     : 38
Average   :
Sum       :
Maximum   :
Minimum   :
Property  :
```

5. In comparison, if you measure the number of Methods, there is a huge difference. $vmview has 100 methods and $vmobject has only 10, as seen in the following screenshot:

```
VMware vSphere PowerCLI 5.8 Release 1                                _  □  x
PowerCLI C:\> $vmview | Get-Member | Where {$_.MemberType -eq "Method"} | Measur
e


Count     : 100
Average   :
Sum       :
Maximum   :
Minimum   :
Property  :


PowerCLI C:\> $vmobject | Get-Member | Where {$_.MemberType -eq "Method"} | Meas
ure


Count     : 10
Average   :
Sum       :
Maximum   :
Minimum   :
Property  :
```

6. Why is there such a big difference in the number of methods since the number of properties is almost identical? If you spend some time going through the list of members for both of the objects, you'll begin to see many methods very similar to the native PowerCLI cmdlets dealing with VM, such as `PowerOnVM`, `PowerOffVM`, `CloneVM`, `RelocateVM`, and so on in the `View` object. For the VM object, you are expected to use the native PowerCLI cmdlets, whereas the View shows the low-level functions that the native PowerCLI are using.

7. The `VMView` retrieved from `Get-View` and stored in `$vmview` is the one that will work in this case. You can use the `Reload()` function and the Runtime property for our needs with inaccessible or invalid VMs. From our `Get-Member` cmdlet, you can see that the `Runtime` Property is an object `VMware.Vim.VirtualMachineRuntimeInfo`. You can explore this object by enumerating it at the command line:

   `$vmview.Runtime`

8. From this enumeration, you will see a property named `ConnectionState`. Unfortunately, all of your VM's are likely "Connected." However, this is the property that will show "inaccessible" or "invalid" if the VM were in that state. So, you will use this property to scope down our target VMs. You can do this using a `Where` statement:

   `$vmview | Where {$_.Runtime.ConnectionState -eq "inaccessible" -or $_.Runtime.ConnectionState -eq "invalid"}`

9. When you run the preceding cmdlet, unless you happen to have a VM in an inaccessible state, there will be no results returned, but you will proceed with the script. The last thing is to use the `Reload()` function. In this case, you will invoke it with `$_.Reload()`:

   `$vmview | Where {$_.Runtime.ConnectionState -eq "inaccessible" -or $_.Runtime.ConnectionState -eq "invalid"} | $_.Reload()`

How it works...

One of the primary reasons to include this recipe is that it is useful to address a real problem you might encounter, but beyond that, it is a fairly simple yet advanced use case for PowerCLI Views which makes it easy to study that topic.

In the recipe, you will use the `Get-Member` cmdlet to examine the objects returned by `Get-View` and `Get-VM` and compare the number of Methods and Properties contained in both. While the number of properties is almost identical, the number of Methods differs significantly. The reason for this is that many of the functions, including the `Get-View` `VMView`, are accomplished with other native cmdlets in PowerCLI.

In this recipe, the work is performed by the `Reload()` function in the VMView. However, you have to scope down the target VMs to be only the VMs that are in the inaccessible or invalid state. You performed that with a simple `Where` statement, however, it required us to go deeper into an object stored in our `VMView` object.

Setting VMware Tool settings from PowerCLI

Over a period of time, VMware Tools have changed within virtual machines. Initially, these tools were configurable within a virtual machine and included drivers for the virtual hardware specific to the operating system. The tools also provided some basic settings to allow the guest virtual machine to synchronize its time with the ESXi host and to automatically upgrade to a newer version of tools on shutdown. Tools also allow graceful shutdown of the guest operating system by executing a script inside the guest, rather than just powering off the VM.

Current versions of VMware Tools provide no user configurable settings inside the guest operating system and now require all of the configurations to occur at the VM level, the same as changing virtual hardware. At the same time, there are no native cmdlets to change the VMware Tool settings in the VM configuration. To make these changes, you will again have to turn to a VMView and custom configuration objects.

Getting Started

To begin this recipe, you will need to open a PowerCLI session and connect to the vCenter sever with our test WinVM1 configured. In addition to the concepts you have already covered, in this recipe you will also need to obtain and pass in a `ChangeVersion` attribute to make the configuration change. This attribute prevents overwriting changes made between the time when a configuration is read and when it is updated. You will need to take the current `ChangeVersion` and pass it back in with the new `VirtualMachineConfigSpec` to update the configuration.

How to do it...

1. The first step is to obtain the `VMView` object for the virtual machine. You can do this using the Get-View cmdlet with the `-VIObject` parameter:

   ```
   $vmview = Get-View -VIObject (Get-VM WinVM1)
   ```

2. The next step is to create a new `VirtualMachineConfigSpec` object:

   ```
   $ConfigSpec = New-Object VMware.Vim.VirtualMachineConfigSpec
   ```

3. Once you have the new object created, you need to take the `ChangeVersion` cmdlet from the view you obtained and place it into our new `ConfigSpec`:

   ```
   $ConfigSpec.ChangeVersion = $vmview.Config.ChangeVersion
   ```

4. Next, you have to create a subkey in the `$ConfigSpec` object with an additional `ToolsConfigInfo` object since this is the configuration you want to change:

   `$ConfigSpec.tools = New-Object VMware.Vim.ToolsConfigInfo`

5. Now, with a `ToolsConfigInfo` object defined in `ConfigSpec`, you can make our configuration changes. The upgrade policy is defined by a `toolsUpgradePolicy` key. You will set the policy to have a value of `upgradeAtPowerCycle`:

   `$ConfigSpec.tools.toolsUpgradePolicy = "upgradeAtPowerCycle"`

6. The other setting change you want to make is to enable the guest time sync with the ESXi host. To do this, you will use the `syncTimeWithHost` key and set it to `$true`. To disable this, you will set the same key to `$false`:

   `$ConfigSpec.tools.syncTimeWithHost = $true`

7. The last step is to issue the reconfiguration function on the View:

   `$vmview.ReconfigVM($ConfigSpec)`

How it works...

This recipe takes you a little deeper into `VMViews`. In this recipe, you create new objects for `VirtualMachineConfigSpec` and `ToolsConfigInfo` and use these objects to make configuration changes by populating some properties in these new objects. Only the properties that you add data to will have any data populated. You also looked at the `ChangeVersion` attribute and used it to populate changes back to a virtual machine through the use of Views. When you invoked the `ReconfigVM()` method, it took the data you populated in the `VirtualMachineConfigSpec` object and changed those properties on the virtual machine. This recipe takes you through a more advanced use case of Views and shows another way that you can use the Methods in the view to invoke a change.

Creating basic reports of VM properties using VMware Tools and PowerCLI

PowerShell has a number of features that make it great for creating reports. It includes native features to export output in CSV and Excel formats to make it easier to work with the retrieved data. You can also search and filter through your cmdlets to easily locate and then scope down the results that you want.

You might need to create basic reports to report properties in the VMs, such as disk free space or virtual hardware defined. Even better, you can be the hero when your supervisor is looking for details about virtual machines related to a particular department or an application if you know how to tap into the potential of PowerCLI and PowerShell's reporting.

As you have seen in the earlier recipes, PowerCLI has cmdlets that look a lot like SQL with `Select`, `Where`, and `Sort` cmdlets. These let you take a result set from a `Get-` cmdlet and quickly return what you're specifically looking for. Beyond these cmdlets, you can also use `-Filter`, `-Name`, `-VM`, or other parameters to return only the specific objects you are looking for.

For this recipe, you will be responding to several requests that you have been assigned. The first is for a list of all the infrastructure virtual machines and the operating systems that they are running. Your manager is concerned with upgrades needed because of the impending end-of-life of an operating system.

The second request from your manager is to create a scheduled task that will generate an e-mail when a virtual machine's disk space has less than 10 percent space free on a filesystem.

Getting Started

To begin this section, open a PowerCLI window and connect to the vCenter server.

How to do it...

1. To begin our first request, we're going to start with a simple `Get-VM` cmdlet to retrieve the list of VM's we're targeting:

   ```
   $InfraVMs = Get-VM -Location "Infrastructure"
   ```

2. The next thing you need to do is look through all of the available fields and see which make sense for your report. You can examine the fields using the `Get-Member` cmdlet:

   ```
   $InfraVMs | Get-Member
   ```

3. From this output, you can select the fields you want. Start with `Name`, `NumCPU`, `MemoryGB`, `Description`, `Host`, and `PowerState`. You can pull only these properties using a simple `Select` cmdlet:

   ```
   $InfraVMs | Select Name, NumCPU, MemoryGB, Description, Host, PowerState
   ```

4. With this, you have the data you want. However, the format isn't the best. How can you get the data out of PowerCLI in a better format? This is where PowerShell's native features come in handy. PowerShell has a cmdlet called `Export-CSV` that saves the data from an object to a CSV file. Even though the cmdlet says CSV, as in comma, you can override the separator and use any other character. In this case, you only need to specify a path for the CSV file using the `-Path` cmdlet:

   ```
   $InfraVMs | Select Name, NumCPU, MemoryGB, Description, Host, PowerState | Export-CSV -Path c:\infravms.csv
   ```

Now, you can open the CSV file with Microsoft Excel or any other spreadsheet software and manipulate it for reporting. PowerShell automatically includes column headings.

5. Comma separated values isn't the only way to get a nice report. PowerShell also has a `ConvertTo-HTML` cmdlet that will format a full HTML file. The unfortunate thing is that the formatting is very plain. It also displays the HTML output to the screen. You will need to use the `ConvertTo-HTML` cmdlet along with `Out-File` to direct the output to a file, unlike the `Export-CSV` cmdlet that does the conversion and saves it in one cmdlet:

```
$InfraVMs | Select Name, NumCPU, MemoryGB, Description, Host,
PowerState | ConvertTo-HTML | Out-File c:\infravms.html
```

6. Fortunately, it is easy to improve the formatting with a little CSS. To do this, you can store a string in a variable and pass it into the `ConvertTo-HTML` cmdlet with the `-Head` parameter:

```
$head = '<style>
body { font-family: Helvetica;
       font-size: 12pt; }
td, th { border: 1px solid black; }
th { color:white;
     background-color:black; }
table, tr, td, th { padding: 4px; margin: 0px }
</style>'

$InfraVMs | Select Name, NumCPU, MemoryGB, Description, Host,
PowerState | ConvertTo-HTML -head $head | Out-File c:\infravms.
html
```

7. The other request you received was to create a script that e-mails when the disk space gets low on a Guest OS filesystem. To comply with this request, you need to work with VMs that have working VMware Tools. Without Tools, you do not have the ability to see into the filesystems or the free space of the filesystems from PowerCLI. First, you will need to save the lines of the script into a text file with the `.ps1` extension.

8. After saving the file, the next step is to ensure that this script knows that it requires the VMware PowerCLI Tools in this script. To do this, you will include an `Add-PSSnapin` cmdlet at the beginning of the script:

```
Add-PSSnapin VMware.VimAutomation.Core
```

9. While you can use `Get-VM` and `Get-HardDisk` as you did in the *Increasing the disk space in a virtual machine* recipe, unfortunately, these cmdlets don't expose the filesystems and their free space either. Once again, we'll be using a View to report the free space. So, the first step is to get a View with all of the Virtual Machines using `Get-View`:

```
$vmview = Get-View -ViewType VirtualMachine
```

10. Next, you want to check the `ToolStatus` parameter, but when you enumerate the View there isn't a property related to VMware Tools; however, there is a Guest parameter populated with an object. Digging into the Guest parameter, you will find a `ToolsStatus` parameter. You can use this to scope down only objects with operating VMware Tools. The two values you want to return are `toolsOk` and `toolsOld`. The second value means VMware Tools are running but are not the latest version. However, these will still work for the needs of this recipe:

```
$VMsWithTools = Get-VM | Get-View | Where {$_.Guest.ToolsStatus
-like "toolsO*"}
```

11. Next, you have an object with all of the virtual machines that you can monitor the disk space for. The next step is to enumerate the disks from these virtual machines. You can run through a quick `ForEach` loop to enumerate all of the values:

```
ForEach ($vm in $VMsWithTools) {

    $vm.Guest.Disk

}
```

12. This is good, but it doesn't quite hit our needs. You have values for the capacity and free space for each filesystem. You can create a computed value for the percentage of free space with a simple math expression. In a Select statement, you can use a hashtable to compute a new property. A hashtable needs to be constructed very specifically, beginning with the @ symbol, followed by a curly brace, and then either `Name=` or `N=` followed by a string that becomes the key name for this property. This section is followed by a semicolon, then either `Expression=` or `E=` and an expression statement. The formatting for the percentage of free space would be constructed as follows:

```
@{N='FreePercent';E={"{0:P1}" -f ( $_.FreeSpace / $_.Capacity ) }}
```

13. Next, you combine this into a select statement with all of the other properties in this object. Now, you will repeat the `ForEach` loop. You will also see a PowerShell formatting string to convert the computed number into a percentage:

```
ForEach ($vm in $VMsWithTools) {

$vm.Guest.Disk | Select *, @{N='FreePercent';E={"{0:P1}" -f (
$_.FreeSpace / $_.Capacity ) }}

}
```

14. However, there is still something missing in the output. You don't know which VM the filesystems belong to. You can solve it with an additional hashtable to bring in the $vm.Name property into this object, where it doesn't exist:

```
ForEach ($vm in $VMsWithTools) {
$vm.Guest.Disk | Select @{N='VM';E={$vm.Name}}, *, @
{N='FreePercent';E={"{0:P1}" -f ( $_.FreeSpace / $_.Capacity ) }}
}
```

15. Finally, you need to scope down the results to just be the ones with under 10 percent free space. That can be done by adding a Where statement:

```
ForEach ($vm in $VMsWithTools) {
$vm.Guest.Disk | Select @{N='VM';E={$vm.Name}}, *, @
{N='FreePercent';E={"{0:P1}" -f ( $_.FreeSpace / $_.Capacity ) }}
| Where { $_.FreePercent * 100 -lt 10}
}
```

16. The output of the preceding ForEach loop is in the listed format. You can easily change that by piping the output to Format-Table:

```
ForEach ($vm in $VMsWithTools) {
$vm.Guest.Disk | Select @{N='VM';E={$vm.Name}}, *, @
{N='FreePercent';E={"{0:P1}" -f ( $_.FreeSpace / $_.Capacity ) }}
| Where { $_.FreePercent * 100 -lt 10} | Format-Table
}
```

17. Next, you need to check for the output and create an e-mail if there are any VMs included with less than 10 percent free space. To do this, you need to modify the ForEach loop to store the output into a variable that adds data during each run through the ForEach. In this example, the output will be stored in a variable called $output:

```
ForEach ($vm in $VMsWithTools) {
$output += $vm.Guest.Disk | Select @{N='VM';E={$vm.Name}}, *, @
{N='FreePercent';E={"{0:P1}" -f ( $_.FreeSpace / $_.Capacity ) }}
| Where { $_.FreePercent * 100 -lt 10} | Format-Table
}
```

18. Now that you have all of the output in a variable, you can measure the variable to see how many rows are contained in it:

```
$outputrows = $output | Measure
```

19. Then, use a simple If statement to see if the number of rows is greater than zero. If they are, you will send an e-mail:

```
if ($outputrows.count > 0) {
```

```
Send-MailMessage -To group@domain.local -Subject "Disk with less
than 10% free disk space" -Body $output -SmtpServer mail.domain.
local -From powershell@domain.local
}
```

20. Now, the PowerShell script can be set up as a scheduled task in Windows, but as it stands, it wouldn't run unless you have set your PowerShell to run unsigned code, which isn't recommended. Therefore, you should sign this PowerShell script using a user certificate, though that goes beyond the scope of this book. See the link in the *See also* section about code signing with PowerShell.

How it works...

Both of the issues solved in this recipe utilize a single PowerShell object to return the results that you are searching for. In the second part of the recipe, the VM name was not included in the object with the other data, so you used a hashtable to include the VM name, which was part of the parent object. You also created a computed hashtable for the percentage of free space. Hashtables are extremely useful when extending the native capabilities of PowerCLI or PowerShell. More than anything else, hashtables look odd from a code standpoint.

In this recipe, you also formatted a number as a percentage and that code looks pretty odd too. While most of PowerShell is very straightforward, there are times when the code looks very foreign and is hard to follow. In the *See also* section, there are links to TechNet articles that will help you go further with hashtables and number formatting.

The primary thing to understand here is that if the data is in a single object, it is very simple to create exports or conversions of the data in common formats such as CSV or any delimited file format and HTML. The HTML formatting can be as simple or elaborate as you want.

There's more...

There is really no limit to the types of reports that you can create from PowerShell. One of the greatest additional features is its ability to export the content you obtain to HTML format. You can add an HTML header and some basic formatting in CSS to the beginning of a PowerShell script so that it can generate an HTML table of the data you have retrieved. This is great for attaching to e-mails or for formatting the body of alert e-mails to make them friendlier for the users receiving them.

See also

▶ Microsoft TechNet, **Formatting Numbers** is available at `http://technet.microsoft.com/en-us/library/ee692795.aspx`

▶ Microsoft TechNet, **Working with Hash Tables** is available at `http://technet.microsoft.com/en-us/library/ee692803.aspx`

▶ TechNet Magazine, **Windows PowerShell: The Many Ways to a Custom Object**, available at `http://technet.microsoft.com/en-us/magazine/hh750381.aspx`

▶ Microsoft TechNet Magazine, **Windows PowerShell**, under **Sign Here, Please**, available at `http://technet.microsoft.com/en-us/magazine/2008.04.powershell.aspx`

4
Working with Datastores and Datastore Clusters

In this chapter, you will cover the following topics:

- ▶ Performing Storage vMotion
- ▶ Finding Raw Disk Mappings in your environment
- ▶ Locating thin or thick provisioned disks
- ▶ Converting thin to thick disks with Storage vMotion
- ▶ Creating and managing datastore clusters
- ▶ Setting Storage DRS automation levels for individual virtual machines
- ▶ Setting Storage DRS automation levels for individual VMs using PowerCLI 6

Introduction

Virtual machines are not static. Virtual workloads change by the minute and vSphere has a lot of features that help administrators handle those dynamic workloads. Log files and growing datasets cause virtual machines to run out of disk space. Fortunately, administrators can easily grow the VMDK files and allocate more disk space to the virtual machine, which was covered in the *Increasing the disk space in a virtual machine* recipe in *Chapter 3, Managing Virtual Machines*.

As several virtual machines increases their disk space, the datastore where they reside might begin to run low on space. Virtual machine snapshots can also constrain the amount of available disk space. Thin provisioned disks in an over-provisioned datastore can completely exhaust the available space. All of these reasons cause administrators to be faced with the manual task of rebalancing virtual machines across datastores.

In early versions of vSphere, the only way to balance datastores was an offline migration of a virtual machine. Since version ESX/ESXi 3.5, VMware added the ability to relocate the storage of a virtual machine online, while the virtual machine is still running. This feature is called **Storage vMotion**. Storage vMotion is an easy task to complete with PowerCLI. A single cmdlet can initiate the relocation of virtual machine files to a new datastore, while the VM remains online.

In vSphere 5, VMware introduced the concept of datastore clusters to help administrators automate the balancing of datastores and fix the manual work. Datastore clusters provide a way to automate datastore balancing by using Storage vMotion.

In addition to the growing disk problems, administrators are faced with the need to convert vSphere virtual disks from thick to thin, and vice versa, in different use cases. Again, this is a task that is well suited for PowerCLI because it is a repetitive task that can be easily done in one quick line.

All of these topics are on the slate in this chapter's recipes.

Performing Storage vMotion

One of the most common things you might need to do with your datastores is to relocate a virtual machine from one datastore to another using Storage vMotion. It allows administrators to rebalance storage utilization across datastores. It also allows administrators to completely vacate a datastore for maintenance or migration. Storage vMotion allows you to nondisruptively move a virtual machine between datastores and borrows its name from vMotion, which allows a VM to relocate from host to host, while the VM remains online.

Getting ready

To begin this recipe, you will need to open a PowerCLI window, connect to a vCenter server, and have a running virtual machine with at least two datastores connected to the host.

How to do it...

To relocate a virtual machine from one datastore to another using Storage vMotion, perform the following steps:

1. In the *Setting up resource pools* recipe in *Chapter 2, Configuring vCenter and Computing Clusters*, you used the `Move-VM` cmdlet to relocate a virtual machine from one resource pool to another. The same cmdlet works for starting a Storage vMotion. The target VM will be the `TTYLinux1` VM imported in *Chapter 3, Managing Virtual Machines*. Lastly, you need to specify a destination with the `-Datastore` parameter:

   ```
   Move-VM -VM TTYLinux1 -Datastore iSCSIDatastore2
   ```

2. While that is simple enough, you certainly do not want to repeat the cmdlet for every VM you might need to move. If every VM on `iSCSIDatastore1` needed to be moved, so that maintenance can be performed on that storage array, or if you need to move VMs onto a new array and decommission an old array, there is an easier way. First, we get all of the VMs from that datastore with the `Get-VM` cmdlet:

   ```
   Get-VM -Datastore iSCSIDatastore1
   ```

3. Next, you will take this cmdlet and pipe that object with all of the VMs into a `Move-VM` cmdlet to initiate the Storage vMotion to the destination datastore. You will see that you do not have to specify a VM name because it's piped into `Move-VM`:

   ```
   Get-VM -Datastore iSCSIDatastore1 | Move-VM -Datastore
   iSCSIDatastore2
   ```

4. Perhaps you don't want to include all of the VMs from a datastore, but maybe all from a particular group, such as all of our `TTYLinux` VMs. You can easily repeat the same cmdlet with the `-VM` parameter and a wildcard match for the VMs you want to move:

   ```
   Get-VM -VM TTYLinux* -Datastore iSCSIDatastore1 | Move-VM
   -Datastore iSCSIDatastore2
   ```

How it works...

The pipeline again saves you from having to type repetitive cmdlets. You can do the work and type a `Move-VM` cmdlet for each VM you want to target, but you don't have to with PowerCLI. You get the benefit of initiating multiple Storage vMotion processes easily. The left-hand side of the pipe retrieves your target VMs that you want to work with, and the right-hand side performs the action.

With any of the `Get-` prefixed cmdlets, you can use wildcard and regex search strings to locate the exact virtual machines that you need. Perhaps the virtual machines you want to move are all in the same folder but they do not necessarily match a pattern in the VM name.

In that case, you can use the -Folder instead of -Datastore and -Name to quickly retrieve the list of VMs to target. It is important to note that we're not using Get-Datastore or Get-Folder here, but Get-VM to return a VM object since you want to move a VM.

Lots of PowerCLI cmdlets can get more complex than this and you can easily pipe objects through many cmdlets, as needed. However, by now, you should certainly be getting a sense of the power of the pipeline.

There's more...

Storage vMotion implies the online relocation of a VM from one datastore to another; however, you can easily move a virtual machine while its offline, too. For some operations, an offline migration is the only option. One of the offline-only migrations is to convert a **Raw Disk Mapping** (**RDM**) into a VMDK. In the next recipe, you will combine several PowerCLI cmdlets to find RDM in your vSphere environment. However, the same Move-VM cmdlet will move a VM whether it is online or offline.

Finding Raw Disk Mappings in your environment

Raw Disk Mappings are a pass through disk type that can be used with virtual machines. With an RDM, the logical disk is connected to the VM directly. Instead of being formatted as a VMFS volume, the logical disk is formatted with an OS filesystem. RDMs allow use cases such as sharing a cluster disk between a physical and virtual node in a cluster, or for taking an existing **Logical Unit Number** (**LUN**) of data and moving it from a physical host to a virtual machine. Some of the use cases are stop-gap uses that are employed during the transition from physical to virtual drives and this might require you to identify and convert RDMs in your environment. PowerCLI is an excellent way to identify virtual machines with RDMs attached.

Getting ready

To begin this recipe, you will need to open a PowerCLI window, connect to a vCenter server, and have a virtual machine with a Raw Disk Mapping. Many environments will not have virtual machines with an RDM, but for those who do, this is handy to use for PowerCLI.

How to do it...

In order to find RDMs in your environment, perform the following steps:

1. The first step is to get all of the VMs in your environment. To do this, issue a Get-VM cmdlet with no parameters, which returns an object with all VMs.

2. Next, you will pipe that object into `Get-HardDisk`. RDMs are attached to a VM, such as VMDK virtual disks, and are listed as hard disks:

   `Get-VM | Get-HardDisk`

3. The object returned actually has a lot of additional data in it. To explore this, use a `Get-Member` cmdlet to explore all of the additional properties available:

   `Get-VM | Get-HardDisk | Get-Member`

 Sometimes, it's more useful to see the data so you can also enumerate the data using a Select * -First 1cmdlet

4. In the list of properties, you will see the `DiskType` property. RDMs are signified by two different disk types: `RawPhysical` and `RawVirtual`. These correspond to the modes for how you can attach an RDM to a VM. You can add the `-DiskType` parameter to the cmdlet and specify the two disk types you want to search for:

   `Get-VM | Get-HardDisk -DiskType "RawPhysical","RawVirtual"`

5. If your goal is to create a report of these disks, you can pipe the output object to a `Select` cmdlet and retrieve the properties you want to display. You can even pipe the `Select` cmdlet to an `Export-CSV` cmdlet, which was covered at the end of *Chapter 3*, *Managing Virtual Machines*, to create a report that can be easily edited.

How it works...

Both of the `Get-` prefixed cmdlets that were used in this recipe are simple and straightforward. The `Get-VM` retrieves all of the VMs in the environment, and the `Get-HardDisk` accepts the VM object that is piped to it and retrieves all of the hard disks associated with each VM. All are returned as objects that can be scoped and reported in whatever format the administrator wants.

There's more...

Locating your RDM might only be half the story. As vSphere has matured, the use cases for using an RDM have decreased and most needs are addressed without needing an RDM. Migrating from an RDM to a VMDK might be the goal, and as mentioned in the *Performing Storage vMotion* recipe, converting an RDM to a VMDK is an offline migration process. Even though it must be performed offline, the conversion invoked from PowerCLI is accomplished with the same `Move-VM` cmdlet that performs a Storage vMotion.

This recipe helped you to identify the RDM in your environment. To convert the RDM to a VMDK, you should shut down the VM and initiate a `Move-VM` cmdlet just as it was done in the *Performing Storage vMotion* recipe. By shutting down the VM first, vSphere will relocate the RDM into a VMDK file that allows you to remove the RDM from vSphere altogether. If you perform this online, the VMDK wrapper file for the RDM will be relocated, but vSphere doesn't convert the RDM.

See also

> ▶ The **Migrating virtual machines with Raw Disk Mappings (RDMs)** page under the **VMware Knowledge Base** title which is available at `http://kb.vmware.com/kb/1005241`

Locating thin or thick provisioned disks

In the early version of ESXi, all VMDKs were thick provisioned disks, which means that all of the data sections of the disk were preallocated onto the backend storage. Thick provisioned disks can be inefficient, especially when there is a large amount of white space or unused space inside of the disk. For instance, if you have a 100 GB disk and only 21 GB is actually used by the guest operating system, you've lost 79 GB of usable disk space in your datastore that could be used by other virtual machines. As storage in vSphere evolved, and as virtualization matured, the concept of thin provisioned disks was introduced in vSphere.

Thin provisioning is the concept of allocating only the data sections of a disk that have data and not allocating any zeroed out sections of the disk. Thin provisioning can save a tremendous amount of backend storage since most virtual machines include some free space. Since the free space is not allocated, the use of thin disks allows administrators to oversubscribe a datastore and allocate more space than that is available in the datastore.

In this recipe, you'll explore how to identify the VMDK disk type and learn which virtual machines have thick provisioned and thin provisioned disks.

Getting ready

For this recipe, you will need to open a PowerCLI window, connect to vCenter, and have a few virtual machines running.

How to do it...

In order to explore and learn which virtual machines have thick provisioned and thin provisioned disks, perform the following steps:

1. Building on the previous recipe, you will begin with a `Get-VM` cmdlet and the `Get-HardDisk` cmdlet in order to find the disks you need to target:

   `Get-VM | Get-HardDisk`

2. If you pipe the object you retrieve to `Get-Member`, you can enumerate the properties available:

   `Get-VM | Get-HardDisk | Get-Member`

3. One of the properties of the object is `StorageFormat`. This is the property you can use to identify either thin or thick disks. Unlike the `DiskType` property, there isn't a parameter to retrieve all of the thin or thick disks. For this, you will need to pipe the object to a `Where` cmdlet:

   `Get-VM | Get-HardDisk | Where {$_.StorageFormat -like "Thin"}`

How it works...

In the previous recipe, you were able to scope the hard disks returned using the `-DiskType` parameter in the `Get-HardDisk` cmdlet. There isn't a `-StorageFormat` parameter so you are forced to use a `Where` cmdlet to trim down the results to only the set that you want. This is less efficient than using native filtering in the prefixed `Get-` cmdlet, but it works just the same. In the end, you have an object with only the hard disks that you're interested in. Now, it is time to use them. You will find their detailed usage in the next recipe.

There's more...

It is important to note that the concept of thin and thick provisioned disks is not unique to VMware and many storage platforms include this capability. Most vendors who perform thin provisioning in the storage platform recommend that you do not use VMware's thin provisioning in addition to the storage vendor's thin provisioning. Thin on thin can lead to major problems as the data grows and can cause unintended downtime if the VMFS datastore or, worse, the array runs out of disk space. There is also a performance penalty to thin provisioning, and high-performance or business critical applications often recommend using thick provisioning. This leads to the need to identify thin provisioned disks and convert them to thick provisioned disks. In this recipe, you will build the PowerCLI to identify the VMware thin provisioned disks, and in the next recipe, you will use this to identify and convert them using Storage vMotion.

Converting thin to thick disks with Storage vMotion

In the previous recipe, you explored how to identify virtual disks that are thin provisioned in vSphere. In this recipe, you will take that concept further. You will take the virtual machines identified as thin provisioned and create PowerCLI cmdlets to convert these using Storage vMotion.

As with all of the capabilities, there are occasions when you need to transition from a thick provisioned disk to a thin provisioned disk, or vice versa. VMware Fault Tolerance requires VMs to have thick provisioned disks. For instance, in this recipe, you'll examine how to convert disks from one disk type to another.

As a part of the conversion, you will be faced with a choice of two types of thick provisioned disks. The two choices are Eager Zero and Lazy Zero thick provisioned disks. Lazy Zero is the default flat file format of VMDK that has always existed in vSphere. Eager Zero is a new type of thick provisioned file that zeroes out the disk area for the new VMDK file being created. By zeroing out the space, any deleted data that might have existed on the disk will be lost. For both of these formats, all of the space required for the disk is required at the time of creation.

Getting ready

To begin this recipe, you will need a PowerCLI window and an active connection to a vCenter host. You will also need a VM that is in the thin provisioned format to be converted.

How to do it...

In order to explore how to convert disks from one disk type to another, perform the following steps:

1. The first step is to pull forward the code to identify the thin provisioned disks in vSphere that was created in the previous recipe:

   ```
   Get-VM | Get-HardDisk | Where {$_.StorageFormat -like "Thin"}
   ```

2. The next step is to perform a `Move-VM` cmdlet. The problem is that the preceding code returns a disk object and not a VM object, but the parent property does specify the VM name. To retrieve VM objects, you will need to rely on the `Get-VM` cmdlet. Explore the VM object's properties with `Get-Member`:

   ```
   Get-VM | Get-Member
   ```

3. You will see that the VM object has a `HardDisks` property and it is the same type of object that `Get-HardDisk` returns. You know that `StorageFormat` contains the data you need to compare with the `HardDisk` object from our previous recipe. Since you need the VM object, and the VM object contains the `HardDisks` property with the `HardDisk` objects, you can use `$_.HardDisks.StorageFormat` as follows:

```
Get-VM | Where {$_.HardDisks.StorageFormat -like "Thin"}
```

4. You now have a VM object returned that can be piped to `Move-VM` to perform Storage vMotion and covert the disks:

```
Get-VM | Where {$_.HardDisks.StorageFormat -like "Thin"} | Move-VM
-Datastore "iSCSIDatastore2" -DiskStorageFormat "Thick"
```

5. If the VM is already on `iSCSIDatastore2`, then the conversion will not work. To make sure that this works properly, you should add an additional statement to the `Where` clause to omit VMs on the destination datastore. You can do this by adding a `Get-Datastore` cmdlet in the `Where` clause and omitting anything on `iSCSIDatastore2`. Using `Get-Datastore` requires you to specify the related object, which is the current object represented by `$_`:

```
Get-VM | Where {$_.HardDisks.StorageFormat -like "Thin" -and (Get-
Datastore -RelatedObject $_) -notlike "iSCSIDatastore2"} | Move-VM
-Datastore "iSCSIDatastore2" -DiskStorageFormat "Thick"
```

6. Lastly, rerun this to find and convert any thin provisioned disks on `iSCSIDatastore2` by moving those to iSCSIDatastore1:

```
Get-VM | Where {$_.HardDisks.StorageFormat -like "Thin" -and (Get-
Datastore -RelatedObject $_) -notlike "iSCSIDatastore1"} | Move-VM
-Datastore "iSCSIDatastore1" -DiskStorageFormat "Thick"
```

How it works...

This recipe works by using Storage vMotion to move a virtual machine from one datastore to another, and by transforming the type of disk used under the VM during the relocation. Storage vMotion includes the capability to change the virtual disk's storage format during a relocation. However, you should also note that you must move from one datastore to another. If you send the cmdlet to relocate the VM to the datastore it already resides on, no transformation will occur.

The `Where` clause in this recipe gets a little more complex. Since you want to omit any VM on the destination datastore, you use the `Get-Datastore` cmdlet. However, this part of the where statement doesn't use data in the current object being piped. Instead, the `Get-Datastore` cmdlet receives the current object as the input and then retrieves its datastore. `Get-Datastore` returns a datastore object and then you can compare it against that result to see whether the VM resides on our destination datastore. Not everything in a Where statement needs to be compared against data in the current object. You can easily run additional cmdlets in parenthesis if you need additional data.

There's more...

You can also convert a thick disk to a thin disk using the same approach in this recipe, by just reversing the storage format you're searching for and the storage format of your destination. Again, if you try to perform a Storage vMotion and disk format change from the same datastore to itself, no change will happen.

Creating and managing datastore clusters

vSphere 5 introduced the new concept of a datastore cluster, which is a storage pool of VMFS datastores clustered in a similar way to how vSphere clusters hosts to share compute. Datastore clusters use Storage DRS to manage the pool of storage.

Storage DRS, as it is known, makes recommendations to balance utilization and workload across datastores in the cluster. Using Storage vMotion, a datastore cluster can balance the utilization of datastores so that the virtual machines are less likely to run out of disk space on their VMFS datastores or experience performance issues due to latency. Storage DRS can either make recommendations or can automatically apply recommendations to move a VM from one datastore to another if a datastore runs low on the available disk space. Storage DRS can also move around virtual machines for the initial placement of a VM that requires more disk space than what is available on any single datastore in a datastore cluster.

Getting ready

To begin this recipe, you will need a PowerCLI window and an active connection to vCenter. You will also need two or more datastores that can be put into a datastore cluster.

How to do it...

In order to start creating datastore clusters and manage them efficiently, perform the following steps:

1. First, create a datastore cluster with the `New-DatastoreCluster` cmdlet. You will need to specify a name for the cluster with the `-Name` parameter. You must also specify the datacenter to place the new datastore cluster using the `-Location` parameter. For simplicity, in this example, you will only specify a name:

   ```
   New-DatastoreCluster -Name "iSCSIcluster" -Location (Get-
   Datacenter Primary)
   ```

2. Next, you need to move some datastores into the datastore cluster. Since you are working with datastores, the first thing is to get your datastore(s). In this example, you will use `Get-Datastore` to get all iSCSI datastores:

   ```
   Get-Datastore -Name iSCSI*
   ```

3. Next, pipe the object to the `Move-Datastore` cmdlet and specify the destination using `-Destination`. For the destination, you will use a `Get-DatastoreCluster` cmdlet to move the datastores into the cluster:

```
Get-Datastore -Name iSCSI* | Move-Datastore -Destination (Get-
DatastoreCluster -Name "iSCSICluster")
```

4. Now that your datastore cluster has a datastore in it, you can change the settings on the cluster. By default, Storage DRS is set to manual mode. This means that it will make recommendations about which VMs to migrate between datastores. To change this, you use the `Set-DatastoreCluster` cmdlet with the `-SdrsAutomationLevel` parameter to set the level to `FullyAutomated` or `Manual`:

```
Set-DatastoreCluster -DatastoreCluster "iSCSICluster"
-SdrsAutomationLevel FullyAutomated
```

5. You can change the thresholds within the datastore cluster. The thresholds control when the cluster recommends VM migrations between the datastores. By default, the default threshold for space utilization is 80 percent, but you can change this with the `-SpaceUtilizationThresholdPercent` cmdlet:

```
Set-DatastoreCluster -DatastoreCluster "iSCSICluster"
-SpaceUtilizationThresholdPercent 90
```

6. Lastly, you can set the IO latency threshold and enable (or disable) the IO load balancing within the cluster. By default, IO load balancing is disabled. To set these, you use the `-IOLatencyThresholdMillisecond` and `-IOLoadBalanceEnabled` parameters:

```
Set-DatastoreCluster -DatastoreCluster "iSCSICluster"
-IoLatencyThresholdMillisecond 30 -IOLoadBalanceEnabled $True
```

How it works...

Datastore clusters are meant to ease the management and balancing of datastores. Datastores experience problems due to growing storage requirements with new VMDKs being added, the allocation of new blocks within thin provisioned disks, and the growing size of snapshots. When utilization of a datastore crosses a threshold, Storage DRS will make the recommendation to move a VM from the datastore to rebalance the utilization. If the Storage DRS automation level is automatic, the recommendation is applied immediately. If it is in manual mode, the administrator must apply the recommendation. In the next recipe, you will explore ways to receive and apply these recommendations from PowerCLI.

As you also explored in this recipe, there is an additional type of balancing that is possible and that is IO load balancing. Although it is disabled in a datastore cluster by default, this type of balancing will benefit environments that have multiple storage arrays, or with storage arrays that link LUNs to specific drives and spindles. Traditional, monolithic storage arrays require administrators to define and carve out LUNs with a lot of thought put into the number of spindles to support a given workload. With virtualization, these workloads are dynamic and a noisy neighbor can spell bad performance for other VMs on a LUN.

Arrays that are architected to spread the IO across large pools of disks that include flash caching or all-flash architectures would be less likely to benefit from IO load balancing. In storage arrays such as these, all of the LUNs presented will likely display similar IO latency numbers, so it's less likely to benefit from the environment. However, if the environment has multiple arrays, the benefits can come back into play.

Setting Storage DRS automation levels for individual virtual machines

In the previous recipe, you created a datastore cluster, moved datastores into the cluster, and set the cluster-wide settings. In the recipe, you set the automation level to fully automated. This means that Storage DRS will direct placement and migrations of virtual machines based on space utilization and/or IO thresholds. However, there might be valid times when a virtual machine should not be moved. Perhaps it's a large virtual machine, or a business-critical virtual machine, where you don't want Storage DRS to automatically move the data disks. In these cases, you can override the cluster-wide settings and you can set a per-VM Storage DRS automation setting. However, PowerCLI does not have a native cmdlet to make this configuration change. In this recipe, you will explore how to use views and objects to make the setting change and save it for vSphere 5.5 using PowerCLI 5.5. If you are using vSphere 6 and PowerCLI 6, continue to the next recipe to learn a slightly different method for the newer version.

Getting ready

To begin this recipe, you will need a PowerCLI window, an active connection to a vCenter server, and a datastore cluster configured in the Fully Automated mode.

How to do it...

In order to set the Storage DRS automation level to `FullyAutomated` for individual virtual machines, perform the following steps:

1. First, you should check the current settings of your datastore cluster so that you know what the default cluster settings are that are being applied to your virtual machines. To do this, use the `Get-DatastoreCluster` cmdlet and store the returned object in a variable:

   ```
   $dsc = Get-DatastoreCluster "iSCSICluster"
   ```

2. With the datastore cluster object stored in a variable, output the variable, and select all of the properties to view the contents:

   ```
   $dsc | Select *
   ```

3. In the output, view the properties defined on your virtual machines. Taking it a step further, take a look at the `ExtensionData` property. It shows that it contains a `VMware.Vim.StoragePod` object. To explore it, output this object using dot notation and the property name:

   ```
   $dsc.ExtensionData
   ```

4. Inside this object, your output will show a `PodStorageDrsEntry` property that contains additional objects. In the `PodStorageDrsEntry` object there are four properties. The first one is `StorageDrsConfig` where you will focus on our setting changes. Inside of `StorageDrsConfig`, there are two properties:

   ```
   $dsc.ExtensionData.PodStorageDrsEntry.StorageDrsConfig
   ```

5. Inside the `StorageDrsConfig` object, you will see two properties. The first is `PodConfig` and this contains the detailed settings for `DatastoreCluster`. The other is `VMConfig` and this is the one that contains the individual VM settings related to Storage DRS.

6. The recipe assumes that your datastore cluster is in the `FullyAutomated` mode and that this is applied to all of the virtual machines. However, for this recipe, assume that you want to omit your domain controller from this policy and you never want it to be relocated.

7. You need to start with a View: the `StorageResourceManager` view. Once you create the View, you will not use it again until you are ready to save all of the changes that are made in this script:

   ```
   $storagemgr = Get-View StorageResourceManager
   ```

8. You need to create an entry in `VmConfig` to change the Storage DRS settings for an individual VM. To build this config, first you need to create a new config object using the `New-Object` cmdlet. The type of the object to create is `VMware.Vim.StorageDrsConfigSpec`.

```
$spec = New-Object VMware.Vim.StorageDrsConfigSpec
```

9. For vSphere 5.x, you will need to output all of the VMs and use a Where statement to choose the ones you want to change:

```
$dsc.ExtensionData.PodStorageDrsEntry.StorageDrsConfig.VmConfig |
Where {$_.VM -like "AD01*"}
```

10. Whoops! That didn't return anything. The reason is that the VM property of `VmConfig` contains a reference to the VM but not the VM's name. You need to adapt this to correlate the VM's name to the reference ID. To do this, you can use `Get-VM` with the ID parameter.

```
$dsc.ExtensionData.PodStorageDrsEntry.StorageDrsConfig.VmConfig |
Where {( Get-VM -Id $_.VM) -like "AD01*"}
```

11. Now that it is scoped properly, you can make changes to these objects. To do this, you will pipe the output from the preceding section into a `ForEach` loop:

```
$dsc.ExtensionData.PodStorageDrsEntry.StorageDrsConfig.VmConfig |
Where {( Get-VM -Id $_.VM) -like "AD01*"} | ForEach {
```

12. Inside the loop, you need a new `VMware.Vim.StorageDrsVmConfigSpec` object to store our change to `StorageDrsVmConfigSpec`. There are two major properties in `StorageDrsVmConfigSpec`. The first is `Operation` or the action to do on `StorageDrsVmConfigSpec`, that is defined. The `Operation` property can be add, edit, or remove. The next property is `Info`. This will contain the existing data that is piped into the `ForEach` loop, and it will also contain our changes to be made:

```
$vmconfig = New-Object VMware.Vim.StorageDrsVmConfigSpec

$vmconfig.Operation = "edit"

$vmconfig.Info = $_

$vmconfig.Info.Enabled = $false
```

13. The last step in the `ForEach` loop is to add this `StorageDrsVmConfigSpec` object into `StorageDrsConfigSpec`.

```
$spec.vmConfigSpec += $vmconfig

}
```

14. Using the `ForEach` loop, all of the `StorageDrsVmConfigSpec` object have been added to the `StorageDrsConfigSpec` object. The last step is to use a `ConfigureStorageDrsForPod` method on the `StorageResourceManager` view that you saved to the `$storagemgr` variable to do the operations you specified in `StorageDrsConfigSpec`:

```
$storagemgr.ConfigureStorageDrsForPod($dsc.ExtensionData.
MoRef,$spec,$true)
```

How it works...

This recipe is one of the most advanced so far in the book. It utilizes a number of objects and illustrates all of the different object types that are available within PowerCLI. Underlying any native cmdlets, these objects get utilized and executed to do the work specified through the more simple cmdlets. Any time you are working with custom configuration specifications, it is important to remember that you need a View in order to have methods to apply these specifications once you define them. If you only define them in objects, they will never make a change. The objects defined are just data in variables until you combine them with a View and a method to perform the changes defined.

In the very beginning of the recipe, you explored the `DatastoreCluster` object returned from the `Get-DatastoreCluster` cmdlet. While exploring that object, you observed nested objects that were several levels deep. By traversing the data at each level, you observed the structure that you needed to create or edit individually. Once you created each object, you nested it back in the same way as the `DatastoreCluster` object. You created a `StorageDrsVmConfigInfo` object, stored it in a `StorageDrsVmConfigSpec` object, and then stored it in a `StorageDrsConfigSpec` object. Once you assembled your data with changes defined in them, you used the `ConfigureStorageDrsForPod` method on the `DatastoreCluster` View to make the changes.

This methodology does not just apply to storage. The recipe is written in this way in order to allow you to view the data structure to help understand how the data is assembled so that you can repeat similar operations for other object types and do similar work to the underlying data.

This comes with a strong word of caution, especially for beginner PowerCLI scripters. While there is a lot of potential for scripting great changes, there is a big potential to damage something. PowerCLI and vSphere attempt to keep you in bounds, but there is a good chance you could do some harm also. It is a very good idea to have a test environment to try scripts and things before doing these scripts on a production system.

There's more...

When you were exploring the `PodStorageDrsEntry` object, you focused on the `StorageDrsConfig` property, but you also saw that there were the `Recommendations`, `DrsFault`, and `ActionHistory` properties. These properties are operational properties within the datastore cluster. The `ActionHistory` property contains a list of all the previous recommendations that have been applied. This property is particularly helpful if you want to report what actions have been applied on this datastore cluster. The `DrsFault` property shows any conditions that cannot be fixed due to a rule or other constraints that will not allow a recommendation to be applied. `Recommendations` shows a list of the active recommendations that are waiting to be applied in the datastore cluster. `Recommendations` is also a property to check on a datastore cluster configured in manual mode.

Setting Storage DRS automation levels for individual VMs using PowerCLI 6

While vSphere 6 requires a slightly different method to update Storage DRS automation levels for virtual machines, most of the information in the previous recipe applies to users running vSphere 6 and PowerCLI 6. The primary difference in vSphere 6 is that you will need to build the entire `StorageDrsVmConfigSpec` object from scratch rather than edit an existing item. In this recipe, you will learn to build a complete `StorageDrsVmConfigSpec` object to set the automation levels for Storage DRS.

Getting ready

To begin this recipe, you will need a PowerCLI window, an active connection to vCenter server, and a datastore cluster configured in the `FullyAutomated` mode.

How to do it...

In order to set the automation levels for Storage DRS for individual VMs, perform the following steps:

1. Just like in the vSphere 5.5 Version of this recipe, you will need to first check the current settings of the datastore cluster. To do this, use the `Get-DatastoreCluster` cmdlet. You will want to store the object returned in a variable for use later in the recipe. Using dot notation, you may investigate the values in the `ExtensionData` property, and for Storage DRS, the `PodStorageDRSEntry` property:

   ```
   $dsc = Get-DatastoreCluster "iSCSICluster"
   $dsc.ExtensionData.PodStorageDRSEntry
   ```

2. In the `PodStorageDrsEntry` object, there are four properties. The first one is `StorageDrsConfig`, and you will focus on this property for our setting changes:

   `$dsc.ExtensionData.PodStorageDrsEntry.StorageDrsConfig`

3. Inside the `StorageDrsConfig` object, you will see two properties. The first one is `PodConfig` that contains the detailed settings for `DatastoreCluster`. The other is `VMConfig`, which is the one, that contains the individual VM settings-related to Storage DRS.

4. The recipe assumes that your datastore cluster is in the `FullyAutomated` mode, and this is applied against all the virtual machines. However, for this recipe, assume that you want to omit your domain controller from this policy, and you never want it to be relocated.

5. You need to start with a View, the `StorageResourceManager` view. Once you create the View, you will not use it again until you are ready to save all the changes that are made in this script:

   `$storagemgr = Get-View StorageResourceManager`

6. For vSphere 6, you will need to first get the VMs you want to change the Storage DRS settings for. Use the `Get-VM` cmdlet to select the VMs you want to change:

   `Get-VM -Name "AD01*"`

7. Take and pipe the object with the VM or VMs into a `ForEach` loop:

   `Get-VM -Name "AD01*" | ForEach {`

8. Next, build a new `StorageDrsVmConfigSpec` object and populate data into it. At this point, we specify an operation to add the new spec. We create a new `ConfigInfo` object, populate it with the VM's reference ID, and set the `Enabled` parameter to `false`. These settings are very similar to what would be set in vSphere 5.5 in the earlier set of steps:

   ```
   $vmconfig = New-Object VMware.Vim.StorageDrsVmConfigSpec
   $vmconfig.Operation = "add"
   $info = New-Object VMware.Vim.StorageDrsVmConfigInfo
   $info.Vm = $_.Id
   $info.Enabled = $false
   $vmconfig.Info = $info
   $spec.vmConfigSpec += $vmconfig
   }
   ```

9. The last step is to use a `ConfigureStorageDrsForPod` method on the `StorageResourceManager` view that you saved to the `$storagemgr` variable in order to do the operations that you specified in `StorageDrsConfigSpec`:

   ```
   $storagemgr.ConfigureStorageDrsForPod($dsc.ExtensionData.
   MoRef,$spec,$true)
   ```

How it works...

In this recipe, you can build your own `StorageDRSConfigSpec` object using references to the VM objects and action operations that are predefined on the `StorageDRSVmConfigSpec` PowerShell object. Unlike the vSphere 5.5 recipe, the vSphere 6 recipe relies on you to create the objects with the operations in them rather than modifying an object that you retrieve from the current configuration. In many ways, this is a much simpler method to implement and maintain because you can select the items to be changed, define the change in the `ConfigSpec` objects, and then run the changes.

5
Creating and Managing Snapshots

In this chapter, you will cover the following topics:

- ▶ Creating a snapshot
- ▶ Getting a list of snapshots in the environment
- ▶ Manipulating the list of snapshots to get better information
- ▶ Scoping and filtering a list of snapshots
- ▶ Removing targeted snapshots
- ▶ Finding lost or unknown snapshots
- ▶ Creating a function to automatically remediate snapshots
- ▶ Scheduling automatic snapshot remediation
- ▶ Creating a snapshot management module

Introduction

Snapshots are one of the best features in vSphere. Snapshots are the safety net built into the platform that allows you to easily revert to a previous known good state in the event that something happens in a VM. Some environments run regularly scheduled snapshots. Some administrators use them only before changes are made in the environment.

Over a period of time, virtual machine snapshots grow to a point that they can exhaust all of the available disk space on a datastore. Administrators can take multiple snapshots and each snapshot references a parent, which means that there is added overhead for IO since the system has to combine multiple files in order to find the correct data to return. The extra latency is minimal and the benefits of the snapshot far outweigh the time.

Snapshots do, however, introduce a management point, since you should not let snapshots linger for too long for fear of exhausting disk space. PowerCLI is an easy way to script and maintain your snapshots and ensure that they do not cause problems in the environment. Since you're dealing with many snapshots and many VMs, PowerCLI can handle management with short and simple cmdlets.

In this chapter, you will start with simple cmdlets to create, report, and remove snapshots. You will move on toward more complex topics such as creating a function to automatically manage snapshots based on specific criteria. You will then take that function and turn it into an easily distributable module that other administrators or users might be able to reuse.

Creating a snapshot

There are lots of reasons why you might want to create a snapshot, and like many other processes, PowerCLI really shines when you need to create more than one at the same time. Creating a snapshot is an easy process from a native cmdlet. What is even better is that the cmdlet can accept piped input of virtual machines that allows you to quickly create snapshots for groups of servers. Many times, when deploying updates or patches to software, you need to create snapshots on multiple servers running the same application. PowerCLI is perfect for the job.

There are two types of snapshots. You can take snapshots that include the memory to return the VM to a running state, including the memory at the time that it is taken. This type of snapshot allows a VM to be brought back to a running state with an active application. There are also snapshots that simply snap the disk, but these would make a crash-consistent version of the virtual machine since the running application might have data in memory that has not been preserved. Reverting to this type of snapshot would boot the virtual server as if it had been reset while running or like it had crashed.

Getting Started

To begin this recipe, you will need to open a PowerCLI window and you should have an active connection to vCenter server. In *Chapter 3*, *Managing Virtual Machines*, you imported a virtual appliance named `SliTaz4.0`, but an alternative virtual appliance was mentioned and named `TTYLinux`. For this recipe, you will import a copy of `TTYLinux` and use it for many of the recipes moving forward in the book.

`TTYLinux` is one of the smallest virtual appliances and it is freely distributed, making it a great lab virtual machine. The link to download this is in the *See Also* section of this recipe. Once you have downloaded the virtual appliance and unzipped the files, import it with the command such as changing the source location and name of your ESXi host:

```
Import-vApp -Source C:\va\TTYLinux\TTYLinux.ovf -Name "TTYLinux1" -VMHost
esxsrv1.domain.local.domain.local
```

To start the VM after importing it, you will use the `Start-VM` cmdlet:

```
Start-VM -VM "TTYLinux1"
```

How to do it...

In order to create a snapshot using a native cmdlet of PowerCLI, perform the following steps:

1. To create a new snapshot, you will use the `New-Snapshot` cmdlet. It requires two parameters, `-VM` for the VM to be targeted and `-Name` for the name of the snapshot:

   ```
   New-Snapshot -VM TTYLinux1 -Name "My First Snapshot"
   ```

2. In larger environments, you might not know the exact name of a VM, so sometimes, it is easier to first use a `Get-VM` cmdlet to find the VMs you want to target and then pipe them into the `New-Snapshot` cmdlet. This is also effective if you have multiple VMs with similar names that all need to be snapped:

   ```
   Get-VM -Name TTY*

   Get-VM -Name TTY* | New-Snapshot -Name "My Second Snapshot"
   ```

3. Creating a snapshot with only the required parameters creates a `PoweredOff` snapshot of the disk, which is crash-consistent. This means that any data in memory might not be preserved in the snapshot and the VM would think that it started from a crash state on the next boot. However, you can also include a snapshot on the RAM for the VM to revert to a powered on state. To do this, add the `-Memory` parameter:

   ```
   Get-VM -Name TTY* | New-Snapshot -Name "Before Maintenance"
   -Memory
   ```

4. You can also create a snapshot that quiesces the disk. This means that it temporarily stops all writes so that it can take a snapshot of the disk knowing that there is no activity in progress. To do this, add the `-Quiesce` parameter:

   ```
   Get-VM -Name TTY* | New-Snapshot -Name "After Maintenance" -Memory
   -Quiesce
   ```

5. The other important parameter to add is a description. The description can be any string of characters. It can be used to add notes about the snapshot, such as who requested it or a date when it can be safely removed:

   ```
   Get-VM -Name TTY* | New-Snapshot -Name "After Installation"
   -Memory -Quiesce -Description "Requested by John in Accounting"
   ```

How it works...

The `New-Snapshot` cmdlet is a single purpose, native cmdlet for creating snapshots on vSphere from PowerCLI. The cmdlet accepts just a few parameters, and the recipe walks you through an explanation of those parameters from just the required parameters to using all of them in a single command.

The `New-Snapshot` cmdlet directs the creation of a snapshot within vSphere. What's important to point out is that although the vSphere Windows client and the vSphere Web Client both take a snapshot with memory, by default, the PowerCLI cmdlet takes the snapshot without memory unless you include the `-Memory` parameter.

The default `New-Snapshot` cmdlet also doesn't quiesce the disk unless the parameter is sent. Quiescing the disk ensures that no partial writes are captured with the snapshot and it ensures better data consistency. In a Windows VM, the quiescence process calls the Windows **Volume Shadow Copy Services** (**VSS**) to ensure that the IO is paused long enough for the snapshot to be taken and then IO is resumed. VSS coordinates not only with the Windows operating system, but it can also coordinate with applications running in the VM to take better point-in-time backups.

There's more...

While VMware Tools in Microsoft Windows uses VSS to coordinate, not all applications might be supported by VSS. This leads to more crash-consistent snapshots and backups. It is possible to extend the functionality to third-party applications by stopping the application or calling a utility to stop IO using the scripts shipped as a part of the VMware Tools. VMware has knowledge base articles on the topic and there are many blog posts that chronicle how to take better snapshots to ensure less data loss.

See also

- **Understanding virtual machine snapshots in VMware ESXi and ESX (1015180)** on the VMware Knowledge Base page is available at `http://kb.vmware.com/kb/1015180`

- **Application quiescing with Windows 2008 R2 SP1 and Windows 2012 with vSphere Data Protection, VMware Data Recovery, and third-party backup software (2044169)** on the VMware Knowledge Base page is available at `http://kb.vmware.com/kb/2044169`

- Mike Laverick's blog on OVF/OVA downloads with TTYLinux OVF is available at `http://www.mikelaverick.com/download/`

Getting a list of snapshots in the environment

Now that you've created a bunch of snapshots in your environment, it is time to keep track of them. Creating a list of snapshots is extremely easy. This recipe will cover how to get that list and perform basic manipulation for reporting on your snapshots.

Getting Started

To begin this recipe, you will need a PowerCLI window and an active connection to the vCenter host. You will also need to have VMs with snapshots, either created as in the previous recipe or created by some other method.

How to do it...

In order to get the created list of snapshots and perform basic manipulation for reporting on these snapshots, perform the following steps:

1. The logical cmdlet to retrieve a list of snapshot objects is the `Get-Snapshot` cmdlet. You can assume that you can run it with no parameters, such as `Get-VM`, and return a list of all the snapshots in the environment. If you execute the following cmdlet on its own, you get an error:

   ```
   Get-Snapshot
   ```

2. The error states that a mandatory object VM is not found. This means that the cmdlet is expecting a VM object to be passed with a list of targeted VMs that should be checked for snapshot. To list all of the snapshots in the environment, you must first run a `Get-VM` cmdlet and then pipe it to `Get-Snapshot`:

   ```
   Get-VM | Get-Snapshot
   ```

   The output for the preceding command is shown in the following screenshot:

   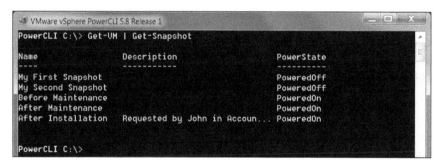

3. The default output from the `Get-Snapshot` cmdlet does not show you which VMs these correspond to. You need to add a column to the default output to make it more usable. To do this, you need to pipe the output to a `Select` statement and add the VM column to the output:

   ```
   Get-VM | Get-Snapshot | Select VM, Name, Description, PowerState
   ```

4. Perhaps you are searching for a particular VM, or group of VMs, instead of all the VMs in a large environment. You can easily add parameters to `Get-VM`, the same parameters that you used in the previous chapters. For example, use the `-Name` parameter to just return our `TTYLinux1` VM:

```
Get-VM -Name TTY* | Get-Snapshot | Select VM, Name, Description,
PowerState
```

How it works...

`Get-Snapshot` requires a VM object to be passed. If you need to check for snapshots on a particular VM, you can pass the VM by its name into the cmdlet and PowerCLI will go and find the VM. However, you'll most likely pipe in the output of a `Get-VM` cmdlet into `Get-Snapshot`. In the recipe, you worked with this method and returned a VM object that contains all of the VMs of the environment and then ran `Get-Snapshot` against all of the VMs.

At the end of the recipe, you looked at scoping down the result of VMs using the `-Name` parameter in the `Get-VM` cmdlet. You should use the `-Name` parameter in this case rather than another method, such as using a `Where` statement, to perform the scoping because it saves some time. In more complex scripts and very large environments, scoping in the cmdlet rather than using a `Where` statement improves the runtime of the script. Both will certainly work, and in this example, there is a negligible difference in the time required. There are more ways than these to accomplish your task, so it becomes a personal preference how you approach it.

There's more...

Even though you can use parameters on `Get-VM` to scope your list of targeted VMs to check for snapshots, you can string together additional cmdlets and pipe their output to `Get-VM`. Again, it comes back to a question of preference and efficiency. You can run a `Get-VMHost` cmdlet and pipe that into `Get-VM` or `Get-Datacenter` and pipe the output into `Get-VM`. However, you can easily use `-Location` and have `Get-VM` retrieve a list based on these same named objects in vSphere without the need of a separate cmdlet.

There are some occasions when you will want to scope based on the data returned in the `Snapshot` object. In those cases, you have no choice but to scope using a `Where` statement, and you will examine those examples in the next recipe in this chapter.

Manipulating the list of snapshots to get better information

Wouldn't it be great if a list could tell you how large your snapshots are growing? Wouldn't it help if you are running low on disk space to know which snapshot is the largest in your environment? By default, the consumed amount of disk space is not a property that gets returned to the list of output for `Get-Snapshot`. However, you can easily add properties to the View of the data being returned. This is a native PowerShell function intended to build on the default views and return any property contained in the objects returned by a cmdlet. Although you will examine it when used with snapshots, the same technique works for any PowerShell object.

Getting Started

To begin this recipe, you will need a PowerCLI window and an active connection to vCenter. You will also need one or more virtual machines with snapshots present.

How to do it...

In order to manipulate the list of snapshots to get better information about them, perform the following steps:

1. As you observed in the previous recipe, the information returned by default from the `Get-Snapshot` cmdlet doesn't tell you all that you want to know. One handy column that can be leveraged is the `Created` column. The `Created` column contains the timestamp of when the snapshot was created. To return this, you simply need to use a `Select` command with the column included:

   Get-VM | Get-Snapshot | Select VM, Name, Created

2. The size columns are additional handy columns that you might need to report. For this recipe, you will use the `SizeGB` column (`SizeMB` is another available column) and then sort it based on this value so that it orders snapshots from the largest to smallest in size:

   Get-VM | Get-Snapshot | Select VM, Name, SizeGB | Sort SizeGB - Desc

The output for the preceding command is shown in the following screenshot:

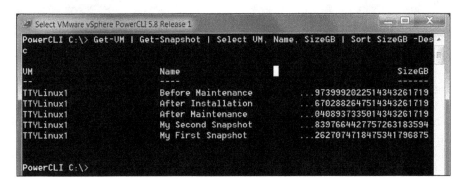

3. In the preceding screenshot, you can see that the `SizeGB` column is truncated in the default output because it is too large to fit. Since that is not very usable, you can use a calculated value to make this more human-readable. To do this, you will use the Name/Expression syntax for a calculated value:

   ```
   Get-VM | Get-Snapshot | Select VM, Name, Created, @
   {N="Size";E={"{0:N2} GB" -f ($_.SizeGB)}}
   ```

4. The last step to make this list more useful is to add a sort so that you get the largest snapshots listed first. To do this, you use the `Sort` statement and pass in the name of the column to sort by. By default, the `Sort` statement will be in an ascending order, but to override this, use the `-Desc` parameter:

   ```
   Get-VM | Get-Snapshot | Select VM, Name, Created, @
   {N="Size";E={"{0:N2} GB" -f ($_.SizeGB)}} | Sort SizeGB
   -Desc
   ```

How it works...

In this recipe, you worked with calculated values. You created a property named `Size` and wrote an expression that formats the number to two decimal places. To do this, you used a formatting feature in PowerShell called the `-f` operator. You specified a format of `{0:N2}`, which means the variable will be output as a number with 2 decimal places.

There's more...

The PowerShell `-f` operator is probably one of the most cryptic things that you'll encounter in PowerShell and PowerCLI. It looks completely foreign to the rest of the language, but it works by taking a formatting string in double quotes and applying it to the variable that is specified after the `-f` operator.

You can use it to simply format the output, or you can use it to format and then store the reformatted information back into a variable. There are thousands of blog articles and TechNet articles that explain the -f operator and format manipulation in PowerShell. It goes far beyond the scope of this book, but it's a concept that needs to be introduced. It is very useful for formatting readable output, and especially for creating reports in PowerCLI.

See also

▸ The Microsoft TechNet page, and the **Formatting Numbers** topic, is available at http://technet.microsoft.com/en-us/library/ee692795.aspx

Scoping and filtering a list of snapshots

As with other cmdlets, there are several ways to approach scoping and filtering a list of snapshots in PowerCLI. There are several methods and some will have benefits in one situation over another. In this recipe, you will look at a couple of different ways to scope a list of snapshots.

Getting Started

To begin this recipe, you will need a PowerCLI window and an active connection to vCenter. You should also have completed the previous recipe.

In this recipe, you will write PowerCLI that will scope down the list of snapshots returned in the output object to only include snapshots that were created more than 30 days ago. However, since your environment might not have snapshots more than 30 days old, you can modify the code to work with 30 minute old snapshots.

You will begin by manually inputting a date to compare against the Created column of the snapshot object returned by Get-Snapshot.

How to do it...

In order to approach scoping and filtering a list of snapshots in PowerCLI, perform the following steps:

1. To begin this recipe, you will use the same Get-VM and Get-Snapshot commands you began the previous recipe with. However, this time, you will use a Where cmdlet and input today's date as a string. For this example, 1/1/2014 will be used:

```
Get-VM | Get-Snapshot | Where {$_.Created -LT "1/1/2014"}
```

Make sure that you use a date greater than today's date when running this cmdlet, so that any snapshots created today in the *Creating a snapshot* recipe will appear.

2. While using a string to compare against the `Created` column is great for interactive sessions, if you want to script an automatic removal after 30 days, you will need to compute 30 days prior. To do this, you can use the PowerShell `Get-Date` cmdlet. To create the timestamp for 30 days ago, use the `Get-Date` cmdlet and use the `AddDays` method of its returned object to add `-30` days. Execute this PowerShell and it outputs a date exactly 30 days ago to the second that it was run:

```
(Get-Date).AddDays(-30)
```

3. Combine the `Get-Date` cmdlet with the PowerCLI you previously used, adding it to a `Where` cmdlet that compares `Created` to the timestamp generated by `Get-Date`. You will use the less than, `-LT`, operator for the `Where` statement.

```
Get-VM | Get-Snapshot | Where {$_.Created -LT (Get-Date).
AddDays(-30)} | Select VM, Name, Created
```

4. Unless you have taken a while between the *Creating a snapshot* recipe and this recipe, your snapshots are not 30 days old. To test the functionality, modify the script to change `AddDays` to `AddMinutes`, and you can use `AddMinutes` for the remainder of the recipe. As long as it has been at least 30 minutes since you created your snapshots, this should return the list of snapshots you created:

```
Get-VM | Get-Snapshot | Where {$_.Created -LT (Get-Date).
AddMinutes(-30)} | Select VM, Name, Created
```

5. In your environment, you might also be concerned with snapshots growing to sizes more than 5 GB. (5 GB is an arbitrary number, just for an illustration.) You can combine this case into the `Where` cmdlet. You will compare `SizeGB` to be greater than (`-GT`) and `5`. You might also want to repeat the calculated value from the previous recipe to show the size of the snapshot:

```
Get-VM | Get-Snapshot | Where {$_.Created -LT (Get-Date).
AddMinutes(-30) -AND $_.SizeGB -GT 5} | Select VM, Name, Created,
@{N="Size";E={"{0:N2} GB" -f ($_.SizeGB)}}
```

 Unless you are running this against an active environment, this should not return any snapshots since the TTYLinux1 snapshots are all very small.

How it works...

In this recipe, you took the `Get-Snapshot` object and output only the snapshots that were older than 30 days. Since your test environment likely doesn't have snapshots that are 30 days old, you changed this to test against snapshots that are only 30 minutes old. This should have returned a list of snapshots that you created in the *Creating a snapshot* recipe of this chapter.

The point to take away is that you can use the `Get-Date` cmdlet to return the present time, and then use methods included in the date object to manipulate the date to go back in time 30 days using the `AddDays`, 30 minutes using the `AddMinutes`, or any other amount of time with the provided methods. You can make the `Add` methods subtract by providing them with a negative number.

In this recipe, you didn't scope based on the `Get-` cmdlets, but instead used a `Where` statement. As you have progressed through the chapter, you used the `Get-` cmdlets to scope as it was available, but this is a case where there are no parameters to scope on the `Get-Snapshot`. Instead, you have to examine the methods and properties that are returned by the `Snapshot` object and utilize the `Where` statement to work with those values.

There's more...

Utilizing what you learned in this chapter will allow you to easily target any particular snapshot that you might encounter that needs to be removed or altered. This recipe is also powerful for reporting and creating proactive alerts of your environment once thresholds are crossed. You can take code similar to this and create a scheduled task, a recipe which is covered later in the chapter, and have it e-mail these alerts on a daily basis so that you get a heads up on what snapshots exist and how they have grown in your environment.

Removing targeted snapshots

Sometimes, you will want to remove a specific snapshot, but not all of the snapshots on a virtual machine. Sometimes, you will want to clean up an entire virtual machine. In this recipe, we're going to work with two of the virtual machines that you created snapshots with in the first recipe of the chapter.

Getting Started

From the first recipe of this chapter, you created snapshots onto `TTYLinux1`. In the first part of this recipe, you will remove a single snapshot from the tree of snapshots on this VM. To do this, you will match it against the name of the snapshot that was created.

Once you remove a targeted, single snapshot, you will remove all of the snapshots on the VM, like you would after a project is complete or a software upgrade is completed successfully. The VM will have three generations of snapshots still present and with a single PowerCLI cmdlet you will remove them all.

To begin this recipe, you will need a PowerCLI window and a connection to the vCenter host with TTYLinux1 (or the virtual machine you used in the *Creating a snapshot* recipe).

How to do it...

In order to first remove a single snapshot from the tree of snapshots on a VM, and then all of the snapshots on the VM, perform the following steps:

1. Since you know the name of the VM you are targeting, you will use a Get-Snapshot with the -VM parameter to list all of the snapshots:

    Get-Snapshot -VM TTYLinux1

    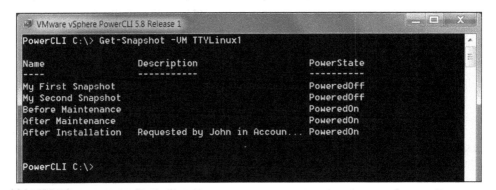

2. In the preceding screenshot, note that the second snapshot has the name, My Second Snapshot. This is the snapshot you want to remove. To do this, you can use the Remove-Snapshot cmdlet, but you need to pass in a snapshot. To get the snapshot to target, use the Get-Snapshot cmdlet with the -Name parameter:

    Get-Snapshot -VM TTYLinux1 -Name "My Second Snapshot" | Remove-Snapshot

3. When you execute the command, it will prompt you to confirm the removal of this single snapshot file. You can suppress this with a -Confirm:$false cmdlet. In all, it is a painless process to remove a specific snapshot.

4. Far easier, you can remove all of the snapshots on a VM with a single and quick cmdlet. You will run a Get-Snapshot -VM cmdlet and pipe that to Remove-Snapshot with a -Confirm:$false cmdlet, and all of the snapshots are instantly and irreversibly removed:

    Get-Snapshot -VM TTYLinux1 | Remove-Snapshot -Confirm:$false

5. Like other PowerCLI, the -VM parameter can accept a wild card input so that you can do this on certain VMs or all VMs in an environment.

6. Going back to the example of removing all snapshots more than 30 days, you can combine the Where statement from the *Scoping and filtering a list of snapshots* recipe into this chapter:

```
Get-Snapshot -VM * | Where {$_.Created -LT (Get-Date).
AddDays(-30)} | Remove-Snapshot
```

 When targeting large groups of VM or snapshots, use the -Confirm:$false cmdlet with caution. It will initiate the removal process and there isn't any going back. If you execute without the -Confirm:$false cmdlet, PowerCLI will prompt you to confirm the removal of each snapshot.

How it works...

In this recipe, you will look at two scenarios. The first scenario is to remove all snapshots on a particular VM. This is easier since you can use the Get-VM cmdlet to retrieve the VM or VMs that you want to target and then you can pipe it to Get-Snapshot and retrieve all of the snapshots on these VMs. Once you have a list of snapshots, you can pipe it to Remove-Snapshot and the snapshots will be removed from the virtual machine. There is no real scoping or difficulty in doing this.

In the second scenario, it follows the same basic pattern, but you don't want to remove all of the snapshots. So, you need to take our Get-Snapshot and scope it with a Where statement to get just the snapshot that you want. There's a lot of data that you can use to target a specific snapshot, but the easiest way is to refer to it by a name or by a date created.

In this recipe, you will work with both scenarios. Using Get-Snapshot and piping it to the Remove-Snapshot cmdlet is the most functional way to remove a snapshot since you don't have to know a lot of specifics about the snapshot you're targeting. Using Get-Snapshot allows you the benefit of exploring before you remove a snapshot.

There's more...

There are many different properties in a snapshot object that you can use to scope and target snapshots for removal. One great aspect of PowerCLI is that it can handle scoping and targeting for any scenario imaginable, that is, as long as the data is there in the snapshot object to interpret or discern which VMs should be included in a removal process. You should spend more time examining the data in a snapshot object using the Get-Member cmdlet and the Select * cmdlets after Get-Snapshot. Take a look at the data points and explore the objects contained inside. PowerCLI can be like onions with layers of depth inside them.

Find lost or unknown snapshots

While not common, there are times when a snapshot might get lost or not be reported properly. Many times, this happens as a result of backup software leveraging snapshots and not cleaning up properly. So what is a lost or unknown snapshot? It's a VM where the VM definition points to a VMDK that has a parent; however, vSphere does not show that it has a snapshot.

What is an easy way to see a VMDK with a parent defined? Snapshots usually take the parent's name and append `-0000#` to the end of the name to create the snapshot disk. In the definition of this VMDK file, it points to a parent. VMware Horizon View uses this method to create **linked clones** where a parent is shared by many child VMs. However, in a vSphere environment, it can spell trouble if the child disk grows to the same size as its parent and it is not intended to be a long term snapshot. Although the problem appears to be more prevalent in the earlier version of vSphere, it is still a great way to dig in and understand a little more about snapshots.

Getting Started

To begin this recipe, you will need a PowerCLI window and a connection to vCenter with one or more VMs present.

How to do it...

In order to find lost or unknown snapshots, perform the following steps:

1. The first step is to get a list of VMs that meet the criteria outlined in the introduction—find VMs with disks that have filenames that end in `0000` and a number. To do this, you use `Get-VM` and pipe it to a `Get-HardDisk` cmdlet to return a list of the actual disk files. Then, pipe that output to a Where statement and look for names that contain `0000`:

   ```
   Get-VM | Get-HardDisk | Where {$_.Filename -like "*0000*"}
   ```

2. The next step is to get a list of VMs with known snapshots. That's easy, since you wrote that command string in earlier recipes. You use `Get-VM` and pipe it to `Get-Snapshot`:

   ```
   Get-VM | Get-Snapshot
   ```

3. Next, you need to compare the two sets of output and look for differences. The problem is that you have a `HardDisk` object and a Snapshot object. You can't compare these, so what do you need to compare? VMs. Modify the hard disks command string and you can select the Parent property, which is the name of the VM. Since you know that the `Snapshot` object has a property named `VM`, create a calculated value named `VM` with the data from Parent:

```
Get-VM | Get-HardDisk | Where {$_.Filename -like "*0000*"} |
Select @{N="VM";E={$_.Parent}}
```

4. When executed, you can see that there is duplicate output in this. To fix that, you can use the Get-Unique cmdlet with an -AsString parameter:

```
Get-VM | Get-HardDisk | Where {$_.Filename -like "*0000*"} |
Select @{N="VM";E={$_.Parent}} | Get-Unique -AsString
```

5. Next, you need to perform a similar Select and Get-Unique on the Snapshot object you returned.

```
Get-VM | Get-Snapshot | Select VM | Get-Unique -AsString
```

6. The last step is to combine these two objects into a Compare-Object cmdlet. You will also want to add a -Property VM so that the compare occurs on the single property:

```
Compare-Object $(Get-VM | Get-HardDisk | Where {$_.Filename -like
"*0000*"} | Select @{N="VM";E={$_.Parent}} | Get-Unique -AsString)
$(Get-VM | Get-Snapshot | Select VM | Get-Unique -AsString)
-Property VM
```

```
VMware vSphere PowerCLI 5.8 Release 1
PowerCLI C:\> Compare-Object $(Get-UM | Get-HardDisk | Where {$_.Filename -like
"*0000*") | Select @(N="UM";E=($_.Parent)) | Get-Unique -AsString) $(Get-UM | Ge
t-Snapshot | Select UM | Get-Unique -AsString) -Property UM

UM                                      SideIndicator
--                                      -------------
TTYLinux2                               <=
TTYLinux3                               <=
Ubuntu14 (2d4a9667-d3e9-4416-b738-3a... <=
TTYLinux (f69c2f56-7728-426d-ae96-b5... <=

PowerCLI C:\> _
```

 If you execute this against your environment and get the output, you have lost or unknown snapshots. In a test environment, it would be unexpected to get the output from running this code. If you do not get an error, then the syntax is correct.

7. Repeat the last line of code and assign it to $Targets variable:

```
$Targets = Compare-Object $(Get-VM | Get-HardDisk | Where {$_.
Filename -like "*0000*"} | Select @{N="VM";E={$_.Parent}} | Get-
Unique -AsString) $(Get-VM | Get-Snapshot | Select VM | Get-Unique
-AsString) -Property VM
```

8. To begin to repair the lost or unknown snapshot, you need to create a new snapshot on the VM. Use the variable you created with the `Compare-Object` cmdlet in a `ForEach` loop and create a new snapshot on each object:

```
ForEach ($Target in $Targets) {
 New-Snapshot -Name "SnapRepair" -VM $Target.VM
}
```

9. In the last step, remove the snapshot you just created:

```
ForEach ($Target in $Targets) {
 Get-Snapshot -VM $Target.VM | Remove-Snapshot
 -Confirm:$false
}
```

10. Rerun the `Compare-Object` cmdlet you created to see whether the VM is still showing a VMDK with the "0000" in the filename and no active snapshot.

How it works...

This recipe gets into a very specific use case for PowerCLI, but it seems to occur with certain backup software and configurations. Later versions of vSphere seem not to be as susceptible to the issue, but the earlier versions of backup software that leveraged the VCB backup method seemed to have some issues with removing snapshots cleanly. However, beyond the snapshot and backup use case, there might be other times when you need to compare the two lists in PowerCLI. This recipe shows you all that you need to know to compare two objects, even if they are of different types and don't match. The output is usable and can be leveraged to do real work based on the differences.

So, what happened in the recipe? The first thing is that you obtained a list of hard disk filenames using the `Get-HardDisk` cmdlet with `Get-VM` piped into it. This gave you a list of matches where their filename pointed to a file with `0000` in the name, indicating that it is some sort of a snapshot. The second part of the compare was a list of known snapshots from the `Get-Snapshot` cmdlet. The differences between the lists would be the unknown snapshots.

The last part of the recipe is simple clean up for the situation, which is to create a known snapshot and remove it. The removal process then cleans up all snapshots on the VM because you did not scope or specify them. A lot of times, when this occurred in the wild, creating a new snapshot made vSphere recognize the full tree of parent disk files and reassemble the list of snapshots to be removed. While it is not 100 percent effective even for the use described, it's a great work around and good study of scoping, limiting results to unique values and comparing the two lists.

Creating a function to automatically remediate snapshots

Creating ad hoc commands is very useful in PowerCLI, but there are times and situations where you find yourself repeatedly doing the same tasks, or where your series of cmdlets becomes long for complex tasks. The next step is to create a PowerCLI function that can be reused and repeated. In the case of snapshot management, the function can be initiated from a single command and will run through your established procedure. This makes it easier to do your work.

You can take and store the function in a `.ps1` file that can be referenced easily. While functions are really useful, they must be defined in each new PowerCLI window, so storing them in a file eases that process.

It is also important to understand scope when working with functions. Each PowerShell or PowerCLI window is a scope in itself. Each function is given its own scope as well, which means that each function can use the same variable names, and changing data in the function's variables does not affect variables outside its scope.

Getting Started

To begin this recipe, you will need a text editor or PowerShell editor.

You can use the native PowerShell ISE or a third-party editor such as Dell Software's PowerGUI.

How to do it...

In order to create a function to automatically remediate snapshots, perform the following steps:

1. The first step is to create a new `.ps1` file using your editor and save it to a folder.

2. The next step is to set up a basic function. To do this, you use the function keyword followed by the name of your function and a set of curly braces:

```
function Remove-30DaySnaps
{
}
```

3. You should begin by adding some notes to the function. The 2011 issue of TechNet Magazine provided a template that you can reuse for the notes in a function. Add the following inside the curly braces:

```
<#
.SYNOPSIS
Describe the function here
.DESCRIPTION
Describe the function in more detail
.EXAMPLE
Give an example of how to use it
.EXAMPLE
Give another example of how to use it
.PARAMETER VM
The VM name to query
.PARAMETER DAYSOLD
The number of days if the snapshot exceeds it is removed
#>
```

 Notes are important and should be included. These notes are used to build the online help when Get-Help is called for your custom function.

4. To make this function work, you should define a list of parameters. In this function, you will want to define a single parameter called VM. Below the notes section, add a param section. The param section defines things such as whether this parameter is required, can it be piped in, and a help message if it is missing to prompt the user for the input. You can also set a default value for a parameter that is not mandatory, such as $DaysOld in the example:

```
param (
  [Parameter(Mandatory=$True,
  ValueFromPipeline=$True,
  ValueFromPipelineByPropertyName=$True,
  HelpMessage='Name:')]
  [string[]] $VM,
  [double] $DaysOld = 30
)
```

5. Next, add a `process()` section with the actual work:

```
process {
  Get-Snapshot -VM $vm | Where {$_.Created -LT
  (Get-Date).AddDays(-$DaysOld)} | Remove-Snapshot
  -Confirm:$false
}
```

6. From the file, you can copy all of your assembled code and paste it into your PowerCLI window. You might have to press Return at the end of the script to execute all that you pasted into the window. There is no output expected if the function passes syntax checks.

 Caution: Do not run this on a production system with snapshots unless you really want to remove any snapshots older than 30 days! *This will remove snapshots.*

7. Now you can call the function using the `Remove-30DaySnaps` function name in PowerCLI. To run it, specify the `-VM *` as the parameter.

How it works...

A PowerShell function is a construct that most of the native PowerCLI cmdlets are written in. Since PowerShell is extensible, the VMware teams have utilized functions and modules in order to build and distribute a very functional set of tools. What you are writing is more specific, but the functionality is very similar to what PowerCLI delivers to the vSphere administrator.

A function begins with two major sections: the params and the process. The process section does the work of the function. The params section defines the input expected and controls how the input comes into the function. In addition, you defined some basic notes about the function.

There's more...

This is an incredibly simple function. Functions are most powerful when they are doing complex PowerShell routines that would be difficult or impossible to manually repeat at the command line. There has to be a sufficient amount of complexity to really make use of the power of functions because of the overhead of defining the parameters, input, and output of the function. Another value is packaging, even a simple routine to distribute among administrators or operators who use your environment.

You can see that you didn't call the `.ps1` file in this recipe. This is because the `.ps1` file is unsigned and is using the default PowerShell execution policy, the `.ps1` file cannot be called. In the next recipe, you will look at how to sign the code and allow execution.

The comment block and its format are also important since this is leveraged by the `Get-Help` cmdlet. The `Get-Help` cmdlet will reference the information in this section in order to build the help output. You can test this by running `Get-Help Remove-30DaySnaps` and see what the output looks like.

See also

▶ TechNet Magazine's **Windows PowerShell: Build a Better Function** page is available at `http://technet.microsoft.com/en-us/magazine/hh360993.aspx`

▶ The Dell Software PowerGUI guide is available at `http://en.community.dell.com/techcenter/powergui/m/`

Scheduling automatic snapshot remediation

With the function in an established form, you can take and execute the `.ps1` file using the Windows Task Scheduler. Creating the scheduled task for a PowerShell requires that you set up some basic things within the `.ps1` file that has your function. You will also want to run the function with a given set of parameters for this scheduled task. All of these things can be added into the file.

In order to use a `.ps1` file, you will need to change the execution policy and sign your code file. Code signing for PowerShell and PowerCLI adds a trusted digital signature to a file, allowing PowerShell to identify the source and trust the file if the signature is trusted. Code signed with a trusted publisher can be run on a Windows machine with a more secure PowerShell execution policy set to `AllSigned` or `RemoteSigned`. Code files that have not been signed require the `Unsigned` execution policy to be set, but this is not secure since even malicious PowerShell could be executed. In order to avoid unneeded risk, you should sign your code files.

For corporate networks with a certificate authority installed, signing your code is as simple as obtaining a user certificate and running a simple cmdlet and this recipe will cover the second scenario.

Getting Started

To begin this recipe, you will need the `.ps1` file that you created in the previous recipe along with a text editor or PowerShell editor. The recipe assumes that there is an Active Directory Certificate Authority installed in the environment.

For the purpose of this recipe, assume that your company has a policy that snapshots shouldn't exist for more than 30 days. You will slightly modify the function you created in the previous recipe in order to accept a parameter called `days`, which you will use for the company policy.

The second part of the recipe is to utilize the Windows Task Scheduler to initiate the PowerCLI commands that you want to execute.

To do this task, you will need your .ps1 file created in the previous recipe and you will need to open the Windows Task Scheduler on your PowerCLI machine.

How to do it...

In order to schedule automatic snapshot remediation, perform the following screenshot:

1. For this recipe, you will leverage the function .ps1 file that you created. Make a copy of this file and add one line to the end that will initiate the function and run it for all VMs:

   ```
   Remove-30DaySnaps -VM *
   ```

2. Save the file with the new line to a new filename, Remove-30DaySnaps.ps1, and save it to a known location. For the steps in the book, you can use C:\Scripts as the location, but any location will work. By adding the preceding line to the .ps1 file, when it is run by the scheduler, the function will run for all VMs in the environment.

3. The next step is to sign the .ps1 file so that it can be executed by PowerShell. The default security configuration of PowerShell prevents all external files from being executed. Check the execution policy of your machine:

   ```
   Get-ExecutionPolicy
   ```

4. If the execution policy is set to Restricted, you will need to change this to RemoteSigned. You will need to launch a PowerCLI or PowerShell window with the **Run as Administrator** option in order to execute the following command, otherwise you will receive an access denied error:

   ```
   Set-ExecutionPolicy -ExecutionPolicy RemoteSigned
   ```

5. The next step is to retrieve a code signing certificate from an internal certificate authority.

Obtaining a code signing certificate is outside the scope of this book, but check out the link in the *See also* section for help in order to obtain a code signing certificate.

If the certificate authority that you retrieve your certificate from is trusted throughout your network, you will be able execute this file anywhere on the network. Outside your network, it would not be trusted and would most likely not be allowed to execute. You can distribute signed code by obtaining a certificate from an externally trusted certificate authority.

6. The next step is to sign the `.ps1` file you've created for this scheduled task. To do this, you run the `Set-AuthenticodeSignature` cmdlet:

```
Set-AuthenticodeSignature C:\Scripts\Remove-30DaySnaps.ps1 @(Get-
Children cert:\\CurrentUsers\My -codesigning)[0]
```

7. To schedule the file to run, you will need to supply a command line into the Task Scheduler. The first part of the command line to supply is the location of the PowerShell executable:

```
C:\Windows\system32\windowspowershell\v1.0\powershell.exe
```

8. PowerShell either requires that you include the snap-in for a language such as PowerCLI to activate all of its cmdlets or you can specify a Console File. From Task Scheduler, the console file is probably the easier method. You specify `-PSConsoleFile` to `powershell.exe`:

```
C:\Windows\system32\windowspowershell\v1.0\powershell.exe
-PSConsoleFIle "c:\Program Files\VMware\Infrastructure\vSphere
PowerCLI\vim.psc1"
```

9. Lastly, you need to specify the location of your custom `.ps1` file to execute. You specify the name of your script file with `& 'path\to\file'`:

```
C:\Windows\system32\windowspowershell\v1.0\powershell.exe
-PSConsoleFIle "c:\Program Files\VMware\Infrastructure\vSphere
PowerCLI\vim.psc1" "& 'C:\scripts\Remove-30DaySnaps.ps1'"
```

Although you are leveraging the function you created in a `.ps1` file, you can simply pass the process block with the `-VM *` and `AddDays(-30)` in a simple `.ps1` file. However, leveraging the function illustrates that you can do much more complex scripts and you had already written it.

How it works...

The Windows Task Scheduler has all of the functionality built into it to execute processes based on different criteria, such as a specified time or every 2 hours. Since it has scheduling capabilities, it makes it the perfect solution to execute PowerShell. Task Scheduler can run the powershell.exe executable, but you need to specify a `.ps1` PowerShell script file to run.

In addition, this is running PowerCLI commands that are a superset of commands created by VMware. In order for those to be accessible, you either need to add `Add-PsSnapin VMware.VimAutomation.Core` into your `.ps1` file, or you need to define `-PSConsoleFile` on the powershell.exe command line. Keep in mind, you are scheduling this to run in Windows cmd.exe and not in PowerShell directly. You are calling the powershell.exe executable, which in turn will execute the PowerShell and PowerCLI commands inside the specified `.ps1` file.

In the *Using PowerShell Native Capabilities to Schedule Scripts* recipe in *Chapter 7, Creating Custom Reports and Notifications for vSphere*, you will look at another method of scheduling PowerCLI and PowerShell to be run on a schedule with a new feature added in PowerShell 3.0.

See also

► The **Sign Here, Please** subtopic under the **Windows PowerShell** topic in the Microsoft TechNet Magazine, dated April 2008, is available at `http://technet.microsoft.com/en-us/magazine/2008.04.powershell.aspx`

► The **Obtaining a Code Signing Certificate and Signing PowerShell Scripts** page is available at `http://www.mikepfeiffer.net/2010/02/obtaining-a-code-signing-certificate-and-signing-powershell-scripts/`

► The *Using PowerShell Native Capabilities to Schedule Scripts* recipe in *Chapter 7, Creating Custom Reports and Notifications for vSphere*

Creating a snapshot management module

While running all of this from a function and using a `.ps1` file works very well, there is a better way. PowerCLI users can actually take the work of a function and create their own module, just like the ones that are used when you use `Import-Module` in PowerShell or PowerCLI. By creating a module, you can locate this in a default `PSModulePath` for PowerShell and you can import it like any vendor supplied modules. You can also distribute this module to end users, help desk staff, or other administrators to ease management. By taking your function or functions to this level, you can gain portability for the code that you've created.

It is important to note that any script file with one or more functions can become a module. Each module should be a unique name, and each module will be a `.psm1` file located in a directory of the same name as the file.

Getting Started

To begin this recipe, you will need a new PowerCLI window that was not used in the previous recipes, and you will need the file you created in the *Creating a function to automatically remediate snapshots* recipe.

How to do it...

In order to create a snapshot management module, perform the following steps:

1. Creating a module requires locating a `.psm1` file in a specific location. The location required is defined in a PowerShell variable. To check for your locations, you can use `Get-Content` to retrieve the location:

   ```
   Get-Content Env:\PSModulePath
   ```

2. The output of the preceding command lists a series of paths on the Windows system separated by semicolons. Each of these paths is a location where PowerShell will look to try and import a module when instructed. Any of these locations can be used for your custom module.

3. The next step is to navigate to one of the paths listed in `PSModulePath`. Open a Windows Explorer window to one of the paths listed.

If you attempt to open the path that begins with `C:\Users\<user>\ Documents`, it's likely the `WindowsPowerShell` directory will not exist. In this event, you can create the `WindowsPowerShell` directory with a `Modules` directory inside it.

If you open the path that begins with `C:\Windows\System32\`, you will see additional modules present, as shown in the following screenshot:

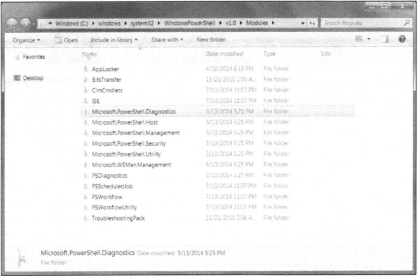

4. Inside the Modules folder, create a new folder named for your custom module. For this example, you can create a module called "30DaySnaps."

5. Open the `.ps1` file you created with the `Remove-30DaySnaps` function inside the file. Save the file as `30DaySnaps.psm1` inside the new `30DaySnaps` folder you created in the `Modules` directory of your choice. No other modifications are needed to your `.ps1` file.

Do not use the `.ps1` file that was used to schedule a recurring task, because that file includes a line to execute the function. You do not want this behavior for a module.

6. In a new PowerCLI window, try to run `Remove-30DaySnaps`.

7. You can see that the error received does not recognize `Remove-30DaySnaps` as the name of a cmdlet, function, or script. Even though the module is in a known path, it has not been loaded, or in PowerShell terminology, it has not been imported. The next step in a new PowerCLI window is to perform an `Import-Module` cmdlet:

   `Import-Module 30DaySnaps`

8. Try again to run `Remove-30DaySnaps` in the PowerCLI window.

9. Now, your function is available. You can run a `Get-Help Remove-30DaySnaps` in the window and you get online help for your custom function.

How it works...

By creating this custom module, which is no more than a PowerShell function in a specially named `.psm1` file, you can distribute the module to other administrators or operators. Each system that needs to be able to run this custom module will locate it in one of the `PSModulePath` locations on their system. Once the `.psm1` file and module's directory structure are located on the system, the user can run an `Import-Module` cmdlet to load the module files, and then the user can invoke any of the functions contained in the module.

Again, in this recipe, code signing is an issue. If the code is not signed with a trusted certificate, users will experience problems loading the module, so the code should be signed with a certificate from a trusted certificate authority. This can be an Active Directory Certificate Authority as long as it is trusted by the systems, which might be controlled by the group policy on the domain.

The names of the functions in your module should follow normal PowerShell naming schemes. There are a set of verbs such as Get, New, Remove, Set, Add, Start, and Stop that are PowerShell recognized verbs. If your module includes nonstandard verbs, PowerShell will display an error on import that your functions do not use the standard verbs which can make them less discoverable. As a general rule, you should stick to the normal verbs to name your functions; however, you are not bound to use these verbs only.

There's more...

The process in our example, the `.psm1` module file, is simple. The modules you write will likely include many functions inside them, and the processes can be much more complex routines or series of commands. One good example is account provisioning for Active Directory and perhaps email. Another example for vSphere might be adapting to the ESXi build routine that you prepared in *Chapter 1, Configuring the Basic Settings of an ESXi Host with PowerCLI*, and turning it into a module and function that you can import and execute with a single function name.

6

Managing Resource Pools, Reservations, and Limits for Virtual Machines

In this chapter, you will cover the following topics:

- ▸ Setting reservations and limits for resource pools
- ▸ Balancing share allocations on resource pools
- ▸ Creating a custom attribute with a number of shares per VM on each resource pool
- ▸ Automating share allocation balancing
- ▸ Reporting shares, reservations, and limits of resource pools and virtual machines

Introduction

The primary concept that vSphere is built around is taking an individual computer, network and disk resources, and combining those into a pool that can be shared by numerous virtual machines. Because of this base concept, resource pools in vSphere are an important concept to understand and administer.

In this chapter, you will look at the basic management of resource pools using PowerCLI. Resource pools have several settings that determine how virtual machines are given access to the available resources. In an environment where there is no contention, virtual machines can consume all of the resources that they request. However, as your environments grow, contention for resources develops, and resource pools are used to set priorities and limits on how much a virtual machine can consume.

Reservations are settings for virtual machines so they are guaranteed a certain amount of CPU or memory regardless of how much contention or slowdown it might create on other virtual machines. **Limits** within resource pools establish a ceiling, where a virtual machine can consume no more CPU or memory. Finally, **Shares** are an allocation allotment that are set at a pool level and then split between individual virtual machines in the pool. Individual virtual machines can have their shares adjusted to give them a priority above or below other virtual machines in the same pool.

These three types of settings are the management points for resource pools. If misconfigured, these settings can cause your environment, or some virtual machines in your environment to behave poorly.

Setting reservations and limits for resource pools

At a conceptual level, reservations and limits make a lot of sense to a vSphere administrator. Reservations are the guarantees to a VM that it will receive at least the specified amount of CPU or RAM as defined by the reservation. Limits are at the other end and limit the VM to not use more than a certain amount of CPU or RAM.

Reservations and Limits can be set on individual VMs, or they can be set on the resource pool and then used by the VMs inside the pool.

Getting ready

To begin this recipe, you will need a PowerCLI window and an active connection to vCenter.

How to do it...

To set reservations and limits for resource pools, perform the following steps:

1. To set resource configurations on one or more virtual machines, you can use the `Set-VMResourceConfiguration` cmdlet. The `Set-VMResourceConfiguration` cmdlet expects to have a resource configuration piped into it, so you must use a `Get-VMResourceConfiguration` cmdlet to retrieve and pipe in an existing configuration:

```
Get-VMResourceConfiguration TTYLinux1
```

2. The next step is to determine the reservation or limit you wish to set. For this example, we will guarantee 25 MB of RAM reservation for `TTYLinux1` and you will set a CPU limit of 1 GHz on the same VM:

```
Get-VMResourceConfiguration TTYLinux1 | Set-
VMResourceConfiguration -MemReservationMB 25 -CpuLimitMhz 1024
```

 All of the other settings on a `Set-VMResourceConfiguration` cmdlet work in the same way as these two examples. Refer to the `Get-Help` for the cmdlet or the online help from VMware for details on all of the parameters.

3. To set the same type of configuration at a resource pool level, you will use the `Set-ResourcePool` cmdlet. Unlike `Set-VMResourceConfiguration`, the `Set-ResourcePool` cmdlet has a parameter to accept the name of the resource pool as a string and does not expect an object. In the next example, you will set a reservation for 1 GB of RAM on the `Production` resource pool:

```
Set-ResourcePool -ResourcePool (Get-ResourcePool -Name Production)
-MemReservationGb 1
```

4. If you have several resource pools named `Production`, the preceding cmdlet will just set multiple resource pools with the configuration. You will see this in the output from the cmdlet. If you need to scope things down to a particular cluster, one option is to use a `Get-ResourcePool` cmdlet inside the `-ResourcePool` parameter, and use the `-Location` parameter to set a specific location as the cluster, `BigCluster`:

```
Set-ResourcePool -ResourcePool (Get-ResourcePool -Location
BigCluster -Name Production) -MemReservationGB 1
```

5. You will see that you used `MemReservationGB`, but there is also `MemReservationMB` available to use. Both set the same parameter, but each represents the value in different measurements. Both can be used interchangeably. To rewrite the previous cmdlet in MB, just change the parameter and value:

```
Set-ResourcePool -ResourcePool (Get-ResourcePool -Location
BigCluster -Name Production) -MemReservationMB 1024
```

6. For the CPU, there is only an MHz representation. To set the CPU reservation on the resource pool, change the parameter to `-CpuReservationMhz` and add a value of `512` to this:

```
Set-ResourcePool -ResourcePool (Get-ResourcePool -Location
BigCluster -Name Production) -CpuReservationMhz 512
```

7. Limits work just the same from a functional standpoint as Reservations. To set a 2 GHz limit and a 4 GB limit on the pool, you will use the `-CpuLimitMhz` and `-MemLimitGB` parameters:

```
Set-ResourcePool -ResourcePool (Get-ResourcePool
-Location BigCluster -Name Production) -CpuLimitMhz 2048
-MemLimitGB 4
```

How it works...

Reservations on virtual machines are a recommendation for business critical applications, ensuring that these virtual machines are allocated all of the requested RAM and CPU that they require. Other times, reservations make sense for latency sensitive applications such as **Voice over IP** (**VoIP**) or transactional applications.

By setting a reservation on the virtual machine, it sets aside that amount of processor or RAM allocation to ensure that the virtual machine receives that amount every time it needs it. No other virtual machine shares these allocations. By setting the limit, you can ensure that the virtual machine never uses more than the capped limit. One use case for a limit on a VM is for small environments with low CPU counts. You might want to limit a VM's CPU in that case so that it doesn't steal the CPU cycles from the ESXi host and make it become unresponsive and unmanageable.

Using the `Set-VMResourceConfiguration` cmdlet, you saw how to set these on a VM basis. You also worked with setting these on a resource pool with the `Set-ResourcePool` cmdlet with the `-CpuLimitMhz`, `-CpuReservationMhz`, `-MemLimitMB`, and `-MemReservationMB` parameters.

Reservations and limits on the pool work in a slightly different way. Once the reservation is made, it's shared among the individual virtual machines in the pool instead of against a single virtual machine.

There's more...

The cmdlets that are used here for setting reservations and limits are also the same cmdlets that are used to set shares on both VM and resource pools. You will explore the cmdlets and their uses with shares in the next recipe.

Although it's not explored in this chapter, VMs can also have other resource limits, such as disk shares and IO limits, set on them. Disk shares work just like CPU and memory shares and are configured with similar cmdlets. The limits for the disk are set with IO per second, so it's a different metric than the arbitrary CPU and memory share values. The disk shares and IO per second limits are not available with resource pools.

See also

► VMware's **vSphere PowerCLI Reference** documentation for `Set-VMResourceConfiguration` is available at `https://www.vmware.com/support/developer/PowerCLI/PowerCLI55/html/Set-VMResourceConfiguration.html`

Balancing share allocations on resource pools

While limits and reservations make immediate sense to most administrators, shares are a more abstract concept, and although they have a major impact in the way that your workloads run on vSphere, these often get overlooked. Many administrators set these on resource pools and forget them later, but with dynamic workloads, the number of virtual machines in a resource pool can affect the distribution of shares and have negative effects on performance.

vSphere environments combine multiple classes of virtual machines. You might have first class, business class, and coach passengers if you'll allow an airline comparison. While you want your first class or mission critical virtual machines, to have all of the resources that they request, business class still get perks, but only after the first class virtual machines needs are met. Coach gets whatever is left over. Your development and test machines can be your coach passengers in vSphere.

If resource pools are left with default allocations, you might not get the desired service level for your VM. For instance, your `Production` pool might have been allocated a share value of `High` that has a default value of 8000. Your `Development` pool might have been allocated a value of `Low` with a default value of 2000. If you have 10 production VMs and only two development VMs, your production VMs will receive 800 shares each and your development VMs will receive 4000 shares each. In essence, your coach passengers are getting better treatment than the first class passengers.

One of the methods to achieve the desired performance is to compute the number of shares for the pool based on the number of virtual machines in the pool. Chris Wahl has an excellent blog post that is the basis for this recipe on his blog at `http://www.wahlnetwork.com`. In his blog post, Chris proposes that you take the number of VMs in a pool and multiply this by the number of shares you want per VM. Each resource pool can have a different number of shares per VM based on the importance of the pool. The computed number becomes the allocation of shares for your pool. This is the method you will use in this recipe.

This method means that you should periodically reallocate shares as the number of virtual machines change in the pool in order to avoid a disproportionate number of shares being allocated to your less important workloads.

Using the default Low, Normal, and High settings is an unoptimized way to allocate your shares. By performing the share allocation based on the number of running VMs, as well as the value of shares per VM, you will ensure the outcome that you want by distributing the shares equitably.

Getting ready

To begin this recipe, you will need a PowerCLI window and an active connection to vCenter. You should also have a couple of resource pools defined with VMs in each resource pool.

How to do it...

In order to balance share allocations on resource pools, perform the following steps:

1. First things first, you will get a list of target resource pools with the Get-ResourcePool cmdlet.

2. The next step is to retrieve a single resource pool into a variable using Get-ResourcePool:

    ```
    $Pool = Get-ResourcePool Production
    ```

3. Once you have a single pool, you will need the total number of VMs in the pool. One way to get this is the Extension data property of the resource pool, $Pool. ExtensionData.Vm.count. Another way to do this is with Get-VM | Measure. You can also limit the count to running VMs by adding a Where statement with the PowerState property:

    ```
    $VmCount = $Pool | Get-VM | Where {$_.PowerState -eq "PoweredOn"} | Measure
    ```

4. The next step is to compute your share values. To do this, simply multiply the number of VMs in $vmcount by the desired share count for CPU and RAM:

    ```
    $PoolShares = $VmCount.Count * 100
    ```

5. The last step is to set the shares for this resource pool. This is handled with the Set-ResourcePool cmdlet. You will need to provide the -ResourcePool parameter, along with the following: -CpuSharesLevel, -NumCpuShares, -RamSharesLevel, and -NumRamShares, in the cmdlet, like you used when creating a resource pool in the *Setting up resource pools* recipe in *Chapter 2, Configuring vCenter and Computing Clusters*:

    ```
    Set-ResourcePool -ResourcePool $Pool.Name
    -CpuSharesLevel:Custom -NumCpuShares $PoolShares
    -MemSharesLevel:Custom -NumMemShares $PoolShares
    -Confirm:$false
    ```

6. You will need to repeat the code again for your additional resource pools. In our example, the environment that was created in *Chapter 2, Configuring vCenter and Computing Clusters*, you had one additional resource pool named `Development`:

```
$Pool = Get-ResourcePool Development

$VmCount = $Pool | Get-VM | Where {$_.PowerState -eq "PoweredOn"}
| Measure

$PoolShares = $VmCount.Count * 50

Set-ResourcePool -ResourcePool $Pool.Name
-CpuSharesLevel:Custom -NumCpuShares $PoolShares
-MemSharesLevel:Custom -NumMemShares $PoolShares
-Confirm:$false
```

How it works...

Adding three simple lines of PowerCLI and your resource pools will have an equitable number of shares as defined by the number of shares per VM that you want in the environment. It is simple multiplication, but as you can see, it's repetitive and would be a good example to use in a `ForEach` loop. Later in this chapter, you will build on this code and reuse it in a large example that also leverages a `ForEach` loop to assign a shares value to multiple resource pools.

There's more...

Chris Wahl's blog post provides sample code for an interactive routine where you are prompted for the number of shares per VM for each resource pool. The script iterates through the list of resource pools in order to configure the shares for each. The link for the blog post is in the *See also* section of this recipe. It provides an additional step that you will find useful since it lets you define and change the share allocation easily.

Moving beyond the interactive method, the next couple of recipes set up a way to create a completely automated routine of updating the share values that can be scheduled or run interactively. You will define the number of shares per VM on each resource pool and the script can run from there.

In addition to Chris Wahl's post, Duncan Epping also posted on the topic back in 2010 on Yellow Bricks. The script and information he presented is also highly applicable to environments today. The script that he provides on his site was written by a colleague. It uses a different method, but it is a great script to explore.

Even more, you can set share allocations on individual VMs, making one more important than another in the resource pool. The VMs take a default allocation of Normal in a resource pool, but this can be overridden to Low, High, or Custom, and can have a custom share value allocated. This is done with the same `Set-VMResourceConfiguration` cmdlet that you used in the previous recipe.

See also

> ▶ Wahl Network: **Understanding Resource Pools in VMware vSphere** is available at `http://wahlnetwork.com/2012/02/01/understanding-resource-pools-in-vmware-vsphere/`

> ▶ Yellow Bricks: **Custom shares on a Resource Pool, scripted** is available at `http://www.yellow-bricks.com/2010/02/24/custom-shares-on-a-resource-pools-scripted/`

Creating a custom attribute with a number of shares per VM on each resource pool

Creating your base script to set the number of shares per resource pool is the first step to fully automating a process to update your share values in order to keep up with the changes and movements in the vSphere environment. Many vSphere environments can have multiple clusters and many more resource pools than just a `Development`, `Test`, and `Production` pool. For example, your environment can have a pool for mission critical systems and one for database systems.

If you have a complex environment, one method to automate the share value assignment is to define a number of shares per VM on each resource pool. This will allow you to take the predefined number of shares and do the computations from the previous recipe in a more automated way. This recipe will walk you through the process of creating a custom attribute and assigning a value to that attribute on each of the resource pools.

In this example, you will use the resource pools created in the *Setting up resource pools* recipe in *Chapter 2, Configuring vCenter and Computing Clusters*. These resource pools are `Production` and `Development`. For this recipe, you will assign `100` as the `SharesPerVM` value on `Production` and 50 as the `SharesPerVM` value on `Development`.

Getting ready

To begin this process, you will need a PowerCLI window, an active connection to vCenter, and Production and Development resource pools created on your cluster.

How to do it...

In order to create a custom attribute with a number of shares per VM on each resource pool, perform the following steps:

1. The first step is to create the custom attribute that can be used on resource pools. The GUI does not display the custom attribute anywhere, but it is visible if you perform a `Get-ResourcePool | Select *` cmdlet from PowerCLI. You will see the `CustomFields` property in the returned values. To begin, examine the existing custom attributes defined in vCenter with `Get-CustomAttributes`:

   `Get-CustomAttribute| Select *`

2. You can see in the output that one of the properties is `TargetType`. The `TargetType` property is a predefined list of objects in vSphere that can have custom attributes defined. To do this, you will use a `TargetType` property of ResourcePool. To create the custom attribute, you will use the New-CustomAttribute cmdlet with a -Name parameter to define the name:

   `New-CustomAttribute -Name SharesPerVM -TargetType ResourcePool`

3. If you perform another `Get-CustomAttribute`, you will see the newly created attribute.

4. Next, you prepare to assign a number to the `SharesPerVM` value on each resource pool. To get a list of resource pools, use the `Get-ResourcePool` cmdlet. Also, run a `Get-ResourcePool | Select *` cmdlet and you can see that `SharesPerVM` is now listed as `CustomField` on each resource pool.

 This is a one-time assignment, but it is something that needs to be done for any new resource pools created in the environment.

5. For each of the resource pools, you will need to repeat the following cmdlet. To assign a value, use the `Set-Annotation` cmdlet. `Set-Annotation` requires `-Entity` that points to an object in vSphere, in this case, it's our resource pool. You must also define `-CustomAttribute` and `-Value` in each cmdlet:

   `Set-Annotation -Entity (Get-ResourcePool Production)`
   `-CustomAttribute SharesPerVM -Value 100`

   `Set-Annotation -Entity (Get-ResourcePool Development)`
   `-CustomAttribute SharesPerVM -Value 50`

6. Run `Get-ResourcePool | Select *` to check the values and you can see that they are set in the `CustomField` property, as shown in the following screenshot:

```
VMware vSphere PowerCLI 5.8 Release 1                              _  □  X

Client                      : VMware.VimAutomation.ViCore.Impl.V1.VimClient

ParentId                    : ResourcePool-resgroup-41
Parent                      : Resources
CpuSharesLevel              : Custom
NumCpuShares                : 50
CpuReservationMHz           : 0
CpuExpandableReservation    : True
CpuLimitMHz                 : -1
MemSharesLevel              : Custom
NumMemShares                : 50
MemReservationMB            : 0
MemReservationGB            : 0
MemExpandableReservation    : True
MemLimitMB                  : -1
MemLimitGB                  : -1
Name                        : Development
CustomFields                : {[SharesPerVM, 50]}
ExtensionData               : VMware.Vim.ResourcePool
Id                          : ResourcePool-resgroup-159
Uid                         : /VIServer=vsphere.local\administrator@192.168.0.240:
                              443/ResourcePool=ResourcePool-resgroup-159/
Client                      : VMware.VimAutomation.ViCore.Impl.V1.VimClient


PowerCLI C:\>
```

How it works...

Creating the custom attribute and assigning values sets you up to be able to create a repeatable and customizable solution for setting share values on your resource pools. Custom attributes can be used on many types of objects in vSphere, but are most commonly defined on VMs. There are third-party software solutions that rely on custom attributes to track values or data that the software needs to operate with vSphere.

In the next recipe, you will take the *Balancing share allocations on resource pools* recipe and extend it into a fully-automated solution that maintains balance for your share settings.

Automating share allocation balancing

In this recipe, you will take the *Balancing share allocations on resource pools* recipe and extend it into a fully-automated solution that maintains the balance of shares in your resource pools. This solution uses the SharesPerVM custom attribute that you created in the previous recipe to compute the number of shares needed for each resource pool. The intent is to create a fully-automated script that can be scheduled and run in order to ensure that your share settings are properly balanced for the desired allocations you want.

Getting ready

To begin this recipe, you will need a PowerCLI window and an active connection to vCenter. You will need the code you created in the *Balancing share allocations on resource pools* recipe, and you need to have completed the *Creating a custom attribute with a number of shares per VM on each resource pool* recipe.

To make this script as multipurpose as possible, you will need to extend your scope slightly. Since vCenter can have multiple clusters, you need to account for multiple clusters and loop through your code in a `ForEach` loop for each cluster. Each cluster will have its own resource pools, and the allocation should be based on the number of VMs inside the cluster and its resource pools only, and not on the total vSphere environment. You should get the number of VMs from each cluster and each resource pool, and you should ensure that you will only compute based on the number of powered on VMs.

To make this as useful as possible, you can also format this as a function and save it as a module in a module location like you learned about in the *Creating a snapshot management module* recipe in *Chapter 5*, *Creating and Managing Snapshots*. This will allow you to drop into PowerCLI, import your custom module, and execute the rebalance anytime that you have made a lot of configuration changes or deployment, before the automated rebalance task runs.

How to do it...

In order to automate the balancing of share allocation so that it's possible to compute the number of shares needed for each resource pool, perform the following steps:

1. The first step is to define the new function, along with the code notes for help and the parameters needed. For now, assume that there are no parameters:

```
function Set-ResourcePoolShareValues
{

    <#
    .SYNOPSIS
    Sets the custom share value of each resource pool based
    on the number of virtual machines in the pool

    .DESCRIPTION
    Calculates the number of shares per resource pool as the
    number of virtual machines multiplied by the SharesPerVM
    custom attribute stored on each virtual machine. Prompts
    user to input a SharesPerVM value if missing or emails
    administrators if run as a scheduled task.

    .EXAMPLE
```

```
Set-ResourcePoolShareValues
#>


param (
)


process {
```

2. In the next step, you will retrieve all of the resource pools in the cluster:

```
$Pools = Get-ResourcePool | Where {$_.Name -ne "Resources"}
ForEach ($Pool in $Pools) {
```

3. Inside the loop, you will use a variable called `$SharesPerVM`, but you want to start with it as a null value on each loop so that you don't inadvertently set a value from the loop before on a different pool. You want to ensure that it is fresh each loop:

```
$SharesPerVM = $null
```

4. You are now ready to move the code you created in the *Balancing share allocations on resource pools* recipe into this `ForEach` loop. Add an `If` statement to check and ensure that `$SharesPerVM` is defined and only compute the `$PoolShares` variable and execute the change if `$SharesPerVM` is defined:

```
$VmCount = Get-VM -Location $Pool | Where {$_.PowerState -eq
"PoweredOn"} | Measure


[int]$SharesPerVM = $Pool.CustomFields.Item("SharesPerVM")


If ($SharesPerVM -ne $null) {


$PoolShares = $VmCount.Count * $SharesPerVM


Set-ResourcePool -ResourcePool $Pool
-CpuSharesLevel:Custom -NumCpuShares $PoolShares
-MemSharesLevel:Custom -NumMemShares $PoolShares
-Confirm:$false
```

5. Now, what if the resource pool doesn't have an allocation of `SharesPerVM` defined? First, you should add an e-mail alert to notify an administrator. So, you should add an `Else` block to the `If` statement. Without a lot of explanation, the two objects will create an e-mail message. The first is `Net.Mail.MailMessage` and the other is `Net.Mail.SmtpClient`, which is the SMTP server information. Each has properties that you will populate in order to send the message:

```
} Else {
  $msg = New-Object Net.Mail.MailMessage
  $smtpServer = New-Object Net.Mail.SmtpClient("hostname")
  $msg.From = "fromaddress@yourcompany.com"
  $msg.To.Add("admin@yourcompany.com")
  $msg.Subject = "Set-ResourcePoolShareValues - Missing
  SharesPerVM value for resource pool"
  $msg.Body = "The resource pool $($Pool.Name) does not
  have a SharesPerVM allocation saved on the pool. Please
  run Set-ResourcePoolShareValues interactively to set and
  save a value."
  $smtp.Send($msg)
} <# End If #>
```

6. Now, close out the `ForEach` loop:

```
} <# End ForEach #>
```

7. Lastly, you need to close the process section of the function and close the function itself:

```
} <# End process #>
} <# End function #>
```

8. You can also prompt the user to get the SharesPerVM value and set it on the resource pool if it is not set. To do this, go back to the `[int]$SharesPerVM = $Pool.CustomFields.SharesPerVM` line of code. Wrap this line of code in an `If` statement, and if the value is missing, prompt the user to set the value using the existing code. To get the user input, you will use the Read-Host cmdlet that creates an interactive prompt with the message you define and stores the value that the user inputs into a variable. You will reuse the code in order to set the value on the custom attribute:

```
If ($Pool.CustomFields.Item("SharesPerVM")-eq "") {
  [int]$SharesPerVM = Read-Host "Missing SharePerVM value
    for $($Pool.Name) resource pool. How many shares per
    VM in the $($Pool.Name) resource pool?"
  Set-Annotation -Entity $Pool -CustomAttribute SharesPerVM
    -Value $SharesPerVM
} Else {
  [int]$SharesPerVM = $Pool.CustomFields.Item("SharesPerVM")
} <# End If #>
```

9. Since you've just added the input, running the function as a scheduled task will now hang and wait for an input if `SharesPerVM` is not defined for a pool. In this case, you want to define a parameter called `RunAsTask`. By default, you will assume that the function is run interactively; however, the `RunAsTask` parameter will accept `$true` as the input and let this run as a scheduled task, suppressing any input requests and send an e-mail or another alert to show an error for a pool without the `SharesPerVM` defined. Also, don't forget to add a line to the notes for this parameter:

    ```
    Param (
        [Parameter(Mandatory=$False,
        ValueFromPipeline=$False,
        ValueFromPipelineByPropertyName=$False,
        HelpMessage='Name:')]
        [Boolean] $RunAsTask = $False
    )
    ```

10. Now change the first line of your `If` statement where you will prompt for input and add a check for the `$RunAsTask` parameter. If the value is `$false`, allow the interactive input, and if `$RunAsTask` is `$true`, it will be skipped:

    ```
    If ($Pool.CustomFields.Item("SharesPerVM") -eq "" -and $RunAsTask
    -eq $false) {
    ```

11. You can run the function in PowerCLI or copy and paste it from a text editor. Once you create the function in the PowerCLI window, you can execute it with no parameters:

    ```
    Set-ResourcePoolShareValues
    ```

12. With everything tested and working, the next step is to save the code as `ResourcePoolShareValues.psm1` file and save it into one of the locations defined by the `PSModulePath` environment variable:

    ```
    Get-Content Env:\PSModulePath
    ```

13. Create a folder called `ResourcePoolShareValues` in one of these path locations and then save the `ResourcePoolShareValues.psm1` file in that directory.

14. You will also need to code sign the file. For more information on that process, see the *Scheduling automatic snapshot remediation* recipe in *Chapter 5, Creating and Managing Snapshots*.

How it works...

The recipe works by reading the custom attribute of `SharesPerVM` from each resource pool and computes the share assignment from the number of powered on VMs, multiplied by the number of shares per VM. It does this for every resource pool and every cluster.

The recipe relies on the custom attribute and on the `Set-ResourcePool` cmdlet in order to make the change. Once the custom attributes are all defined, you can run the function in a scheduled task that are two simple lines:

```
Add-PSSnapIn VMware.VimAutomation.Core

Import-Module ResourcePoolShareValues

Connect-VIServer hostname

Set-ResourcePoolShareValues -RunAsTask $true
```

You can always import and run the function interactively. If you receive an e-mail notification that the SharesPerVM value is not set, you can run the cmdlet and add this value interactively, or you can refer to the instructions in the *Creating a custom attribute with a number of shares per VM on each resource pool* recipe and set it manually.

There's more...

It's important to remember that module files can include more than one function. Similar concepts or management areas can easily be grouped together into a module file. In the next recipe, you will extend this module file with additional functionalities.

See also

▸ The *Scheduling automatic snapshot remediation* recipe in *Chapter 5, Creating and Managing Snapshots*

Reporting shares, reservations, and limits of resource pools and virtual machines

Now that you have the shares fully automated and scheduled, you might need to report the values and ensure that everything is good. Since your shares are defined by the number of VMs and the custom attribute value, you can easily create a table-formatted report that includes the set values for memory and CPU shares and compare that against a computed value. This will help you to verify that your function is working in the future. For this recipe, you will write an additional function to add into your module.

Getting ready

This recipe will continue inside the `.psm1` file that you created in the previous recipe. You will need this file, a text editor, a PowerCLI window, and an active connection to vCenter.

How to do it...

In order to report shares, reservations, and limits of resource pools and virtual machines, perform the following steps:

1. The first step is to get a basic function established. Refer to the *Automating share allocation balancing* recipe and grab the function definition, the code notes, the param, and process blocks of the function and copy those to a new function called `Get-ResourcePoolShareValues`.

2. Once you have created the skeleton of the function, edit the notes for this function. Include a note and a description about what the function will do along the lines of, "`This function retrieves the set values for CPU and memory shares along with the SharesPerVM custom attribute, the number of virtual machines running in the pool, and the computed value of shares to verify the settings.`"

3. This function will have no parameters, so the param block can be left blank.

4. In the process block, you will begin the processing by getting back a list of the resource pools. To do this, use the `Get-ResourcePool` function and exclude the overall `Resources` root-level pool:

   ```
   Get-ResourcePool | Where {$_.Name -ne "Resources"}
   ```

5. To this line of code, add a `Select` statement where you will extract the properties that you want to report back. These properties are `Name`, `NumCPUShares`, `NumMemShares`.

   ```
   Get-ResourcePool | Where {$_.Name -ne "Resources"} | Select Name,
   CPUSharesLevel, NumCPUShares, MemSharesLevel, NumMemShares,
   ```

6. In many vSphere environments, there can be multiple `Production` resource pools with the same name, but the cluster name or ESXi hostname will show where this particular `Production` pool is located. To retrieve it, you need to create a named expression using `Get-View` to retrieve the name of the cluster or host based on the `ExtensionData.Owner` property:

   ```
   @{N="Owner";E={(Get-View $_.ExtensionData.Owner).Name}},
   ```

7. You also want to include the `SharesPerVM` value that is a custom attribute. To add this, use a dot notation to access the value of this property. You will need to create this as a named expression:

   ```
   @{N="SharesPerVM";E={$_.CustomFields.Item('SharesPerVM')}},
   ```

8. The next thing we want to include is the count of the number of VMs in each resource pool. The question that arises is, can you access this in `ExtensionData`, or do you need to use a `Get-VM` cmdlet in an expression? The answer to this question is shown in the following command line:

```
@{N="ActiveVMs";E={(Get-VM -Location $_ | Where {$_.PowerState -eq
'PoweredOn'} | Measure).Count}},
```

9. The next thing that you need to do is compute what the share value should be set to so that the administrator can visually compare it with the set value and know if an update is needed:

```
@{N="ComputedShares";E={[int]$_.CustomFields.Item('SharesPerVM') *
[int](Get-VM -Location $_ | Where {$_.PowerState -eq 'PoweredOn'}
| Measure).Count} }
```

10. The last thing is to force the output into a table. You can do this by piping the output to `Format-Table` or to FT as a shortcut. The final process code, all together, should be as follows:

```
Get-ResourcePool | Where {$_.Name -ne "Resources"} | Select Name,
CPUSharesLevel, NumCPUShares, MemSharesLevel, NumMemShares, @
{N="Owner";E={(Get-View $_.ExtensionData.Owner).Name}}, @
{N="SharesPerVM";E={$_.CustomFields.Item('SharesPerVM')}}, @
{N="ActiveVMs";E={(Get-VM -Location $_ | Where {$_.PowerState -eq
'PoweredOn'} | Measure).Count}}, @{N="ComputedShares";E={[int]$_.
CustomFields.Item('SharesPerVM') * [int](Get-VM -Location $_ |
Where {$_.PowerState -eq 'PoweredOn'} | Measure).Count} } | FT
```

11. Close your process block and close the function, and save the file. You are now ready to test the module:

```
} <# End process #>
} <# End function #>
```

12. Just to be safe, start a new PowerCLI window and connect to vCenter. Perform an `Import-Module ResourcePoolShareValues` function and then run the `Get-ResourcePoolShareValues` function to see the output, as illustrated in the following screenshot:

How it works...

This function pulls a predefined set of properties from the resource pool and computes several additional attributes in order to provide the administrator with the values set and what the computed value of shares should be. This is really just defining a shortcut, since the code to accomplish this is a one-line PowerCLI command. However, with multiple computed values in named expressions, it's not one that you would easily write off the top of your head over and over.

Building on what you learned in the last chapter, you will see how functions can group together functionality around the same management idea.

There's more...

There are no limits to this style of `Get-` function. If you have different reporting views for different staff, but need to generate them often, a function is a handy way to create these. You can create a different function for Bob's view and Tom's view, saving time and effort on your part.

You can take these functions a level deeper and integrate the e-mail code to take the output and send it directly to users. Adding a parameter or writing a modified version of this script will allow you to quickly send a list to a manager who is requesting it without ever having to think about the PowerCLI code again.

If the user prefers to get the report output in a format that they can manipulate further, such as Excel, combine the output with either `Export-CSV` or `ConvertTo-CSV` and then attach it to the e-mail. There are so many combinations and it's surprisingly easy to set these up for all your needs.

See also

▶ The *Sending output to CSV and HTML* recipe in *Chapter 7, Creating Custom Reports and Notifications for vSphere*

7

Creating Custom Reports and Notifications for vSphere

In this chapter, you will cover the following topics:

- ▶ Getting alerts from a vSphere environment
- ▶ Basics of formatting output from PowerShell objects
- ▶ Sending output to CSV and HTML
- ▶ Reporting VM objects created during a predefined time period from VI Events object
- ▶ Setting custom properties to add useful context to your virtual machines
- ▶ Using PowerShell native capabilities to schedule scripts

Introduction

This chapter is all about leveraging the information available to you in PowerCLI. As much as any other topic, figuring out how to tap into the data that PowerCLI offers is as important as understanding the cmdlets and syntax of the language. However, once you obtain your data, you will need to alter the formatting and how it's returned to be used. This is something you've been doing to some extent throughout the book so far with `Select` statements. PowerShell, and by extension PowerCLI, offers a big set of ways to control the formatting and the display of information returned by its cmdlets and data objects. You will explore all of these topics with the recipes in this chapter.

Getting alerts from a vSphere environment

Discovering the data available to you is the most difficult thing that you will learn and adopt in PowerCLI after learning the initial cmdlets and syntax. There is a large amount of data available to you through PowerCLI, but there are techniques to extract the data in a way that you can use. The `Get-Member` cmdlet is a great tool for discovering the properties that you can use. Sometimes, just listing the data returned by a cmdlet is enough; however, when the property contains other objects, `Get-Member` can provide context to know that the Alarm property is a **Managed Object Reference** (**MoRef**) data type.

As your returned objects have properties that contain other objects, you can have multiple layers of data available for you to expose using PowerShell dot notation (`$variable.property.property`). The `ExtensionData` property found on most objects has a lot of related data and objects to the primary data. Sometimes, the data found in the property is an object identifier that doesn't mean much to an administrator but represents an object in vSphere. In these cases, the `Get-View` cmdlet can refer to that identifier and return human-readable data.

This recipe will walk you through the methods of accessing data and converting it to usable, human-readable data wherever needed so that you can leverage it in scripts. To explore these methods, we will take a look at vSphere's built-in alert system.

While PowerCLI has native cmdlets to report on the defined alarm states and actions, it doesn't have a native cmdlet to retrieve the triggered alarms on a particular object. To do this, you must get the datacenter, VMhost, VM, and other objects directly and look at data from the `ExtensionData` property.

Getting ready

To begin this recipe, you will need a PowerCLI window and an active connection to vCenter. You should also check the vSphere Web Client or the vSphere Windows Client to see whether you have any active alarms and to know what to expect. If you do not have any active VM alarms, you can simulate an alarm condition using a utility such as `HeavyLoad`. For more information on generating an alarm, see the *There's more...* section of this recipe.

How to do it...

In order to access data and convert it to usable, human-readable data, perform the following steps:

1. The first step is to retrieve all of the VMs on the system. A simple `Get-VM` cmdlet will return all VMs on the vCenter you're connected to.

2. Within the VM object returned by `Get-VM`, one of the properties is `ExtensionData`. This property is an object that contains many additional properties and objects. One of the properties is `TriggeredAlarmState`:

```
Get-VM | Where {$_.ExtensionData.TriggeredAlarmState -ne $null}
```

3. To dig into `TriggeredAlarmState` more, take the output of the previous cmdlet and store it into a variable. This will allow you to enumerate the properties without having to wait for the `Get-VM` cmdlet to run. Add a `Select -First 1` cmdlet to the command string so that only a single object is returned. This will help you look inside without having to deal with multiple VMs in the variable:

```
$alarms = Get-VM | Where {$_.ExtensionData.TriggeredAlarmState -ne
$null} | Select
-First 1
```

4. Now that you have extracted an alarm, how do you get useful data about what type of alarm it is and which vSphere object has a problem? In this case, you have VM objects since you used `Get-VM` to find the alarms. To see what is in the `TriggeredAlarmState` property, output the contents of `TriggeredAlarmState` and pipe it to `Get-Member` or its shortcut `GM`:

```
$alarms.ExtensionData.TriggeredAlarmState | GM
```

The following screenshot shows the output of the preceding command line:

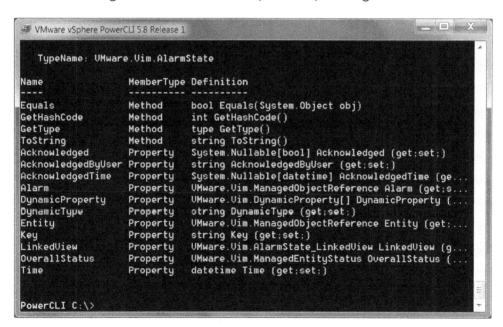

5. List the data in the `$alarms` variable without the `Get-Member` cmdlet appended and view the data in a real alarm. The data returned does tell you the time when the alarm was triggered, the `OverallStatus` property or severity of the alarm, and whether the alarm has been acknowledged by an administrator, who acknowledged it and at what time.

6. You will see that the `Entity` property contains a reference to a virtual machine (as you saw in the *Locating and reloading inaccessible or invalid virtual machines* recipe from *Chapter 3*, *Managing Virtual Machines*). You can use the `Get-View` cmdlet on a reference to a VM, in this case, the `Entity` property, and return the virtual machine name and other properties. You will also see that Alarm is referred to in a similar way and we can extract usable information using `Get-View` also:

    ```
    Get-View $alarms.ExtensionData.TriggeredAlarmState.Entity

    Get-View $alarms.ExtensionData.TriggeredAlarmState.Alarm
    ```

7. You can see how the output from these two views differs. The `Entity` view provides the name of the VM. You don't really need this data since the top-level object contains the VM name, but it's good to understand how to use `Get-View` with an entity.

    On the other hand, the data returned by the `Alarm` view does not show the name or type of the alarm, but it does include an `Info` property. Since this is the most likely property with additional information, you should list its contents. To do so, enclose the `Get-View` cmdlet in parenthesis and then use dot notation to access the `Info` variable:

    ```
    (Get-View $alarms.ExtensionData.TriggeredAlarmState.Alarm).Info
    ```

8. In the output from the `Info` property, you can see that the example alarm in the screenshot is a `Virtual Machine CPU usage` alarm. Your alarm can be different, but it should appear similar to this.

9. After retrieving PowerShell objects that contain the data that you need, the easiest way to return the data is to use calculated expressions. Since the `Get-VM` cmdlet was the source for all lookup data, you will need to use this object with the calculated expressions to display the data. To do this, you will append a `Select` statement after the `Get-VM` and `Where` statement. Notice that you use the same `Get-View` statement, except that you change your variable to `$_`, which is the current object being passed into `Select`:

```
Get-VM | Where {$_.ExtensionData.TriggeredAlarmState -ne $null}
| Select Name, @{N="AlarmName";E={(Get-View $_.ExtensionData.
TriggeredAlarmState.Alarm).Info.Name}},

@{N="AlarmDescription";E={(Get-View $_.ExtensionData.
TriggeredAlarmState.Alarm).Info.Description}}, @
{N="TimeTriggered"; E={$_.ExtensionData.TriggeredAlarmState.
Time}}, @{N="AlarmOverallStatus"; E={$_.ExtensionData.
TriggeredAlarmState. OverallStatus}}
```

How it works...

When the data you really need is several levels below the top-level properties of a data object, you need to use calculated expressions to return these at the top level. There are other techniques where you can build your own object with only the data you want returned, but in a large environment with thousands of objects in vSphere, the method in this recipe will execute faster than looping through many objects to build a custom object. Calculated expressions are extremely powerful since nearly anything can be done with expressions.

More than that, you explored techniques to discover the data you want. Data exploration can provide you with incredible new capabilities. The point is you need to know where the data is and how to pull that data back to the top level.

There's more...

It is likely that your test environment has no alarms and in this case, it might be up to you to create an alarm situation. One of the easiest to control and create is heavy CPU load with a CPU load-testing tool. JAM Software created software named `HeavyLoad` that is a stress-testing tool. This utility can be loaded into any Windows VM on your test systems and can consume all of the available CPU that the VM is configured with. To be safe, configure the VM with a single vCPU and the utility will consume all of the available CPU.

Once you install the utility, go to the **Test Options** menu and you can uncheck the **Stress GPU** option, ensure that **Stress CPU** and **Allocate Memory** are checked. The utility also has shortcut buttons on the Menu bar to allow you to set these options. Click on the **Start** button (which looks like a Play button) and the utility begins to stress the VM.

For users who wish to do the same test, but utilize Linux, StressLinux is a great option. StressLinux is a minimal distribution designed to create high load on an operating system.

See also

▶ You can read more about the HeavyLoad Utility available under the JAM Software page at `http://www.jam-software.com/heavyload/`

▶ You can read more about StressLinux at `http://www.stresslinux.org/sl/`

Basics of formatting output from PowerShell objects

Anything that exists in a PowerShell object can be output as a report, e-mail, or editable file. Formatting the output is a simple task in PowerShell and to some extent you've already been doing some basic formatting in prior recipes. Sometimes, the information you receive in the object is in a long decimal number format, but to make it more readable, you want to truncate the output to just a couple decimal places. You did this in the *Manipulating the list of snapshots to get better information* recipe in *Chapter 5, Creating and Managing Snapshots*, with formatting numbers.

In this recipe, you will take a look at the `Format-Table`, `Format-Wide`, and `Format-List` cmdlets. You will dig into the `Format-Custom` cmdlet and also take a look at the `-f` format operator that you leveraged in *Chapter 5, Creating and Managing Snapshots*, for uses beyond formatting numbers.

The truth is that native cmdlets do a great job returning data using default formatting. When we start changing and adding our own data to the list of properties returned, the formatting can become unoptimized. Even in the returned values of a native cmdlet, some columns might be too narrow to display all of the information.

Getting ready

To begin this recipe, you will need the PowerShell ISE.

How to do it...

In order to format the output from PowerShell objects, perform the following steps:

1. Run `Add-PSSnapIn VMware.VimAutomation.Core` in the PowerShell ISE to initialize a PowerCLI session and bring in the VMware cmdlet. Connect to your vCenter server.

2. Start with a simple object from a `Get-VM` cmdlet. The default output is in a table format. If you pipe the object to `Format-Wide`, it will change the default output into a multicolumn with a single property, just like running a `dir /w` command at the Windows Command Prompt. You can also use `FW`, an alias for `Format-Wide`:

   ```
   Get-VM | Format-Wide
   Get-VM | FW
   ```

3. If you take the same object and pipe it to `Format-Table` or its alias `FT`, you will receive the same output if you use the default output for `Get-VM`:

   ```
   Get-VM
   Get-VM | Format-Table
   ```

4. However, as soon as you begin to select a different order of properties, the default formatting disappears. Select the same four properties and watch the formatting change. The default formatting disappears.

   ```
   Get-VM | Select Name, PowerState, NumCPU, MemoryGB | FT
   ```

5. To restore formatting to table output, you have a few choices. You can change the formatting on the data in the object using the `Select` statement and calculated expressions. This is the method you used in the *Manipulating the list of snapshots to get better information* recipe of *Chapter 5, Creating and Managing Snapshots*. You can also pass formatting through the `Format-Table` cmdlet. While setting formatting in the `Select` statement changes the underlying data, using `Format-Table` doesn't change the data, but only its display. The formatting looks essentially like a calculated expression in a `Select` statement. You provide `Label`, `Expression`, and formatting commands:

   ```
   Get-VM | Select * | FT Name, PowerState, NumCPU, @
   {Label="MemoryGB"; Expression-{$_.MemoryGB}; FormatString="N2";
   Alignment="left"}
   ```

6. If you have data in a number data type, you can convert it into a string using the `ToString()` method on the object. You can try this method on NumCPU:

   ```
   Get-VM | Select * | FT Name, PowerState, @{Label="Num CPUs";
   Expression={($_.NumCpu).ToString()}; Alignment="left"}, @
   {Label="MemoryGB"; Expression={$_.MemoryGB}; FormatString="N2";
   Alignment="left"}
   ```

7. The other method is to format with the `-f` operator, which is basically a .NET derivative. When you used this previously in *Chapter 5, Creating and Managing Snapshots*, there wasn't a lot of explanation. To better understand the formatting and string, the structure is `{<index>[,<alignment>][:<formatString>]}`. Index sets that are a part of the data being passed, will be transformed. The alignment is a numeric value. A positive number will right-align those number of characters. A negative number will left-align those number of characters. The `formatString` parameter is the part that defines the format to apply. In this example, let's take a datastore and compute the percentage of free disk space. The format for percent is `p`:

```
Get-Datastore | Select Name, @{N="FreePercent";E={"{0:p} -f ($_.
FreeSpaceGB / $_.CapacityGB)}}
```

8. To make the FreePercent column 15 characters wide, you add `0,15:p` to the format string:

```
Get-Datastore | Select Name, @{N="FreePercent";E={"{0,15:p} -f
($_.FreeSpaceGB / $_.CapacityGB)}}
```

How it works...

With the `Format-Table`, `Format-List`, and `Format-Wide` cmdlets, you can change the display of data coming from a PowerCLI object. These cmdlets all apply basic transformations without changing the data in the object. This is important to note because once the data is changed, it can prevent you from making changes. For instance, if you take the percentage example, after transforming the `FreePercent` property, it is stored as a string and no longer as a number, which means that you can't reformat it again. Applying a similar transformation from the `Format-Table` cmdlet would not alter your data. This doesn't matter when you're performing a one-liner, but in a more complex script or in a routine, where you might need to not only output the data but also reuse it, changing the data in the object is a big deal.

There's more...

This recipe only begins to tap the full potential of PowerShell's native `-f` format operator. There are hundreds of blog posts about this topic, and there are use cases and examples of how to produce the formatting that you are looking for. The following link also gives you more details about the operator and formatting strings that you can use in your own code.

See also

▶ For more information, refer to the **PowerShell -f Format operator** page available at `http://ss64.com/ps/syntax-f-operator.html`

Sending output to CSV and HTML

On the screen the output is great, but there are many times when you need to share your results with other people. When looking at sharing information, you want to choose a format that is easy to view and interpret. You might also want a format that is easy to manipulate and change.

Comma Separated Values (**CSV**) files allow the user to take the output you generate and use it easily within a spreadsheet software. This allows you the ability to compare the results from vSphere versus internal tracking databases or other systems easily to find differences. It can also be useful to compare against service contracts for physical hosts as examples.

HTML is a great choice for displaying information for reading, but not manipulation. Since e-mails can be in an HTML format, converting the output from PowerCLI (or PowerShell) into an e-mail is an easy way to assemble an e-mail to other areas of the business.

What's even better about these cmdlets is the ease of use. If you have a data object in PowerCLI, all that you need to do is pipe that data object into the `ConvertTo-CSV` or `ConvertTo-HTML` cmdlets and you instantly get the formatted data. You might not be satisfied with the HTML-generated version alone, but like any other HTML, you can transform the look and formatting of the HTML using CSS by adding a header.

In this recipe, you will examine the conversion cmdlets with a simple set of `Get-` cmdlets. You will also take a look at trimming results using the `Select` statements and formatting HTML results with CSS.

This recipe will pull a list of virtual machines and their basic properties to send to a manager who can reconcile it against internal records or system monitoring. It will export to a CSV file that will be attached to the e-mail and you will use the HTML to format a list in an e-mail to send to the manager.

Getting ready

To begin this recipe, you will need to use the PowerShell ISE.

How to do it...

In order to examine the conversion cmdlets using `Get-` cmdlets, trim results using the `Select` statements, and format HTML results with CSS, perform the following steps:

1. Open the PowerShell ISE and run `Add-PSSnapIn VMware.VimAutomation.Core` to initialize a PowerCLI session within the ISE.

2. Again, you will use the `Get-VM` cmdlet as the base for this recipe. The fields that we care about are the name of the VM, the number of CPUs, the amount of memory, and the description:

```
$VMs = Get-VM | Select Name, NumCPU, MemoryGB, Description
```

3. In addition to the top-level data, you also want to provide the IP address, hostname, and the operating system. These are all available from the `ExtensionData.Guest` property:

```
$VMs = Get-VM | Select Name, NumCPU, MemoryGB, Description, @
{N="Hostname";E={$_.ExtensionData.Guest.HostName}}, @
{N="IP";E={$_.ExtensionData.Guest.IPAddress}}, @{N="OS";E={$_.
ExtensionData.Guest.GuestFullName}}
```

4. The next step is to take this data and format it to be sent as an HTML e-mail. Converting the information to HTML is actually easy. Pipe the variable you created with the data into `ConvertTo-HTML` and store in a new variable. You will need to reuse the data to convert it to a CSV file to attach:

```
$HTMLBody = $VMs | ConvertTo-HTML
```

5. If you were to output the contents of `$HTMLBody`, you will see that it is very plain, inheriting the defaults of the browser or e-mail program used to display it. To dress this up, you need to define some basic CSS to add some style for the `<body>`, `<table>`, `<tr>`, `<td>`, and `<th>` tags. You can add this by running the `ConvertTo-HTML` cmdlet again with the `-PreContent` parameter:

```
$css = "<style> body { font-family: Verdana, sans-serif; font-
size: 14px; color: #666; background: #FFF; } table{ width:100%;
border-collapse:collapse; } table td, table th { border:1px solid
#333; padding: 4px; } table th { text-align:left; padding: 4px;
background-color:#BBB; color:#FFF;} </style>"

$HTMLBody = $VMs | ConvertTo-HTML -PreContent $css
```

6. It might also be nice to add the date and time generated to the end of the file. You can use the `-PostContent` parameter to add this:

```
$HTMLBody = $VMs | ConvertTo-HTML -PreContent $css -PostContent
"<div><strong>Generated:</strong> $(Get-Date)</div>"
```

7. Now, you have the HTML body of your message. To take the same data from `$VMs` and save it to a CSV file that you can use, you will need a writable directory, and a good choice is to use your `My Documents` folder. You can obtain this using an environment variable:

```
$tempdir = [environment]::getfolderpath("mydocuments")
```

8. Now that you have a temp directory, you can perform your export. Pipe `$VMs` to `Export-CSV` and specify the path and filename:

```
$VMs | Export-CSV $tempdir\VM_Inventory.csv
```

9. At this point, you are ready to assemble an e-mail and send it along with your attachment. Most of the cmdlets are straightforward. You set up a $msg variable that is a MailMessage object. You create an Attachment object and populate it with your temporary filename and then create an SMTP server with the server name:

```
$msg = New-Object Net.Mail.MailMessage

$attachment = new-object Net.Mail.Attachment("$tempdir\VM_
Inventory.csv")

$smtpServer = New-Object Net.Mail.SmtpClient("hostname")
```

10. You set the From, To, and Subject parameters of the message variable. All of these are set with dot notation on the $msg variable:

```
$msg.From = "fromaddress@yourcompany.com"

$msg.To.Add("admin@yourcompany.com")

$msg.Subject = "Weekly VM Report"
```

11. You set the body you created earlier, as $HTMLBody, but you need to run it through Out-String to convert any other data types to a pure string for e-mailing. This prevents an error where System.String[] appears instead of your content in part of the output:

```
$msg.Body = $HTMLBody | Out-String
```

12. You need to take the attachment and add it to the message:

```
$msg.Attachments.Add($attachment)
```

13. You need to set the message to an HTML format; otherwise, the HTML will be sent as plain text and not displayed as an HTML message:

```
$msg.IsBodyHtml = $true
```

14. Finally, you are ready to send the message using the $smtpServer variable that contains the mail server object. Pass in the $msg variable to the server object using the Send method and it transmits the message via SMTP to the mail server:

```
$smtpServer.Send($msg)
```

15. Don't forget to clean up the temporary CSV file you generated. To do this, use the PowerShell Remove-Item cmdlet that will remove the file from the filesystem. Add a -Confirm parameter to suppress any prompts:

```
Remove-Item $tempdir\VM_Inventory.csv -Confirm:$false
```

How it works...

Most of this recipe relies on native PowerShell and less on the PowerCLI portions of the language. This is the beauty of PowerCLI. Since it is based on PowerShell and only an extension, you lose none of the functions of PowerShell, a very powerful set of commands in its own right.

The `ConvertTo-HTML` cmdlet is very easy to use. It requires no parameters to produce HTML, but the HTML it produces isn't the most legible if you display it. However, a bit of CSS goes a long way to improve the look of the output. Add some colors and style to the table and it becomes a really easy and quick way to format a mail message of data to be sent to a manager on a weekly basis.

The `Export-CSV` cmdlet lets you take the data returned by a cmdlet and convert that into an editable format for use. You can place this onto a file share for use or you can e-mail it along, as you did in this recipe.

In the earlier chapters, you didn't go into much detail on how to create a mail message. This recipe takes you step by step through the process of creating a mail message, formatting it in HTML, and making sure that it's relayed as an HTML message. You also looked at how to attach a file. To send a mail, you define a mail server as an object and store it in a variable for reuse. You create a message object and store it in a variable and then set all of the appropriate configuration on the message. For an attachment, you create a third object and define a file to be attached. That is set as a property on the message object and then finally, the message object is sent using the server object.

There's more...

`ConvertTo-HTML` is just one of four conversion cmdlets in PowerShell. In addition to `ConvertTo-HTML`, you can convert data objects into XML. `ConvertTo-JSON` allows you to convert a data object into an XML format specific for web applications. `ConvertTo-CSV` is identical to `Export-CSV` except that it doesn't save the content immediately to a defined file. If you had a use case to manipulate the CSV before saving it, such as stripping the double quotes or making other alternations to the contents, you can use `ConvertTo-CSV` and then save it to a file at a later point in your script.

Reporting VM objects created during a predefined time period from VI Events object

An important auditing tool in your environment can be a report of when virtual machines were created, cloned, or deleted. Unlike snapshots, that store a created date on the snapshot, virtual machines don't have this property associated with them. Instead, you have to rely on the events log in vSphere to let you know when virtual machines were created.

PowerCLI has the Get-VIEvents cmdlet that allows you to retrieve the last 1,000 events on the vCenter, by default. The cmdlet can accept a parameter to include more than the last 1,000 events. The cmdlet also allows you to specify a start date, and this can allow you to search for things within the past week or the past month.

At a high level, this recipe works the same in both PowerCLI and the vSphere SDK for Perl (VIPerl). They both rely on getting the vSphere events and selecting the specific events that match your criteria. Even though you are looking for VM creation events in this recipe, you will see that the code can be easily adapted to look for many other types of events.

Getting ready

To begin this recipe, you will need a PowerCLI window and an active connection to a vCenter server.

How to do it...

In order to report VM objects created during a predefined time period from VI Events object, perform the following steps:

1. You will use the Get-VIEvent cmdlet to retrieve the VM creation events for this recipe. To begin, get a list of the last 50 events from the vCenter host using the -MaxSamples parameter:

   ```
   Get-VIEvent -MaxSamples 50
   ```

2. If you pipe the output from the preceding cmdlet to Get-Member, you will see that this cmdlet can return a lot of different objects. However, the type of object isn't really what you need to find the VM's created events. Looking through the objects, they all include a GetType() method that returns the type of event. Inside the type, there is a name parameter.

3. Create a calculated expression using `GetType()` and then group it based on this expression, you will get a usable list of events you can search for. This list is also good for tracking the number of events your systems have encountered or generated:

```
Get-VIEvent -MaxSamples 2000 | Select @{N="Type";E={$_.GetType().
Name}} | Group Type
```

4. In the preceding screenshot, you will see that there are `VMClonedEvent`, `VmRemovedEvent`, and `VmCreatedEvent` listed. All of these have to do with creating or removing virtual machines in vSphere. Since you are looking for created events, `VMClonedEvent` and `VmCreatedEvent` are the two needed for this script. Write a `Where` statement to return only these events. To do this, we can use a regular expression with both the event names and the `-match` PowerShell comparison parameter:

```
Get-VIEvent -MaxSamples 2000 | Where {$_.GetType().Name -match
"(VmCreatedEvent|VmClonedEvent)"}
```

5. Next, you want to select just the properties that you want in your output. To do this, add a `Select` statement and you will reuse the calculated expression from Step 3. If you want to return the VM name, which is in a `Vm` property with the type of `VMware. Vim.VVmeventArgument`, you can create a calculated expression to return the VM name. To round out the output, you can include the `FullFormattedMessage`, `CreatedTime`, and `UserName` properties:

```
Get-VIEvent -MaxSamples 2000 | Where {$_.GetType().
Name -match "(VmCreatedEvent|VmClonedEvent)"} | Select @
{N="Type",E={$_.GetType().Name}}, @{N="VMName",E={$_.Vm.Name}},
FullFormattedMessage, CreatedTime, UserName
```

```
VMware vSphere PowerCLI 5.8 Release 1                              _ □ X

PowerCLI C:\> Get-VIEvent -MaxSamples 100000 | Where {$_.GetType().Name -match "
(VmCreatedEvent|VmClonedEvent)"} | Select @{N="Type";E={$_.GetType().Name}}, @{N
="VMName";E={$_.Vm.Name}}, FullFormattedMessage, CreatedTime, Username


Type                 : VmClonedEvent
VMName               : TTYLinux2
FullFormattedMessage : Clone of TTYLinux1 completed
CreatedTime          : 9/13/2014 10:55:43 AM
UserName             : VSPHERE.LOCAL\Administrator

Type                 : VmClonedEvent
VMName               : TTYLinux3
FullFormattedMessage : Clone of TTYLinux1 completed
CreatedTime          : 9/13/2014 10:38:46 AM
UserName             : VSPHERE.LOCAL\Administrator

Type                 : VmCreatedEvent
VMName               : TTYLinux3
FullFormattedMessage : Created virtual machine TTYLinux3 on
```

6. The last thing you will want to do is go back and add a time frame to the `Get-VIEvent` cmdlet. You can do this by specifying the `-Start` parameter along with `(Get-Date).AddMonths(-1)` to return the last month's events:

```
Get-VIEvent -Start (Get-Date).AddMonths(-1) -MaxSamples 2000 |
Where {$_.GetType().Name -match "(VmCreatedEvent|VmClonedEvent)"}
| Select @{N="Type",E={$_.GetType().Name}}, @{N="VMName",E={$_.
Vm.Name}}, FullFormattedMessage, CreatedTime, UserName
```

How it works...

The `Get-VIEvent` cmdlet drives a majority of this recipe, but in this recipe you only scratched the surface of the information you can unearth with `Get-VIEvent`. As you saw in the screenshot, there are so many different types of events that can be reported, queried, and acted upon from the vCenter server.

Once you discover and know which events you are looking for specifically, then it's a matter of scoping down the results with a `Where` statement. Last, you use calculated expressions to pull data that is several levels deep in the returned data object.

One of the primary things employed here is a regular expression used to search for the types of events you were interested in: `VmCreatedEvent` and `VmClonedEvent`. By combining a regular expression with the `-match` operator, you were able to use a quick and very understandable bit of code to find more than one type of object you needed to return.

There's more...

Regular Expressions (**RegEx**) are big topics on their own. These types of searches can match any type of pattern that you can establish or in the case of this recipe, a number of defined values that you are searching for. RegEx are beyond the scope of this book, but they can be a big help anytime you have a pattern you need to search for and match, or perhaps more importantly, replace. You can use the `-replace` operator instead of `–match` to not only to find things that match your pattern, but also change them.

See also

> ▸ For more information on Regular Expressions refer to `http://ss64.com/ps/syntax-regex.html`

> ▸ The **PowerShell.com** page: **Text and Regular Expressions** is available at `http://powershell.com/cs/blogs/ebookv2/archive/2012/03/20/chapter-13-text-and-regular-expressions.aspx`

Setting custom properties to add useful context to your virtual machines

Building on the use case for the `Get-VIEvent` cmdlet, Alan Renouf of VMware's PowerCLI team has a useful script posted on his personal blog (refer to the *See also* section) that helps you pull the created date and the user who created a virtual machine and populate this into a custom attribute. This is a great use for a custom attribute on virtual machines and makes some useful information available that is not normally visible.

This is a process that needs to be run often to pick up details for virtual machines that have been created. Rather than looking specifically at a VM and trying to go back and find its creation date as Alan's script does, in this recipe, you will take a different approach building on the previous recipe and populate the information from the found creation events. Maintenance in this form would be easier by finding creation events for the last week, running the script weekly, and updating the VMs with the data in the object rather than looking for VMs with missing data and searching through all of the events.

This recipe assumes that you are using a Windows system that is joined to AD on the same domain as your vCenter. It also assumes that you have loaded the Remote Server Administration Tools for Windows so that the Active Directory PowerShell modules are available. This is a separate download for Windows 7. The Active Directory Module for PowerShell can be enabled on Windows 7, Windows 8, Windows Server 2008, and Windows Server 2012 in the **Programs and Features** control panel under **Turn Windows features on or off**.

Getting ready

To begin this script, you will need the PowerShell ISE.

How to do it...

I order to set custom properties to add useful context to your virtual machines, perform the following steps:

1. Open the PowerShell ISE and run `Add-PSSnapIn VMware.VimAutomation.Core` to initialize a PowerCLI session within the ISE.

2. The first step is to create a custom attribute in vCenter for the `CreatedBy` and `CreateDate` attributes:

   ```
   New-CustomAttribute -TargetType VirtualMachine -Name CreatedBy
   New-CustomAttribute -TargetType VirtualMachine -Name CreateDate
   ```

3. Before you begin the scripting, you will need to run `ImportSystemModules` to bring in the Active Directory cmdlets that you will use later to lookup the username and reference it back to a display name:

   ```
   ImportSystemModules
   ```

4. Next, you need to locate and pull out all of the creation events with the same code as the *Reporting VM objects created during a predefined time period from VI Events object* recipe. You will assign the events to a variable for processing in a loop in this case; however, you will also want to change the period to 1 week (7 days) instead of 1 month:

   ```
   $Events = Get-VIEvent -Start (Get-Date).AddDays(-7) -MaxSamples
   25000 | Where {$_.GetType().Name -match "(VmCreatedEvent|VmClonedE
   vent)"}
   ```

5. The next step is to begin a `ForEach` loop to pull the data and populate it into a custom attribute:

   ```
   ForEach ($Event in $Events) {
   ```

6. The first thing to do in the loop is to look up the VM referenced in the Event's `Vm` parameter by name using `Get-VM`:

   ```
   $VM = Get-VM -Name $Event.Vm.Name
   ```

7. Next, you can use the `CreatedTime` parameter on the event and set this as a custom attribute on the VM using the `Set-Annotation` cmdlet:

```
$VM | Set-Annotation -CustomAttribute "CreateDate" -Value $Event.
CreatedTime
```

8. Next, you can use the `Username` parameter to lookup the display name of the user account who created the VM using Active Directory cmdlets. For the Active Directory cmdlets to be available, your client system or server needs to have the Microsoft **Remote Server Administration Tools** (**RSAT**) installed to make the Active Directory cmdlets available. The data coming from `$Event.Username` is in DOMAIN\ username format. You need just the username to perform a lookup with `Get-AdUser`, so that you can split on the backslash and return only the second item in the array resulting from the split command. After the lookup, the display name that you will want to use is in the Name property. You can retrieve it with dot notation:

```
$User = (($Event.UserName.split("\"))[1])

$DisplayName = (Get-AdUser $User).Name
```

9. To do this, you need to use a built-in on the event and set this as a custom attribute on the VM using the `Set-Annotation` cmdlet:

```
$VM | Set-Annotation -CustomAttribute "CreatedBy" -Value
$DisplayName
```

10. Finally, close the `ForEach` loop.

```
} <# End ForEach #>
```

How it works...

This recipe works by leveraging the `Get-VIEvent` cmdlet to search for events in the log from the last number of days. In larger environments, you might need to expand the `-MaxSamples` cmdlet well beyond the number in this example. There might be thousands of events per day in larger environments.

The recipe looks through the log and the `Where` statement returns only the creation events. Once you have the object with all of the creation events, you can loop through this and pull out the username of the person who created each virtual machine and the time they were created. Then, you just need to populate the data into the custom attributes created.

There's more...

Combine this script with the next recipe and you have a great solution for scheduling this routine to run on a daily basis. Running it daily would certainly cut down on the number of events you need to process through to find and update the virtual machines that have been created with the information.

You should absolutely go and read Alan Renouf's blog post on which this recipe is based. This primary difference between this recipe and the one Alan presents is the use of native Windows Active Directory PowerShell lookups in this recipe instead of the Quest Active Directory PowerShell cmdlets.

See also

▶ **Virtu-Al.net**: **Who created that VM?** is available at `http://www.virtu-al.net/2010/02/23/who-created-that-vm/`

Using PowerShell native capabilities to schedule scripts

In the previous recipes of this book, you scheduled PowerCLI to run as scheduled tasks by defining and referring to a script file (`.ps1`) and running it from the PowerShell.exe executable. There is potentially a better and easier way to schedule your processes to run from PowerShell and PowerCLI and those are known as Scheduled Jobs. Scheduled Jobs were introduced in PowerShell 3.0 and distributed as part of the Windows Management Framework 3.0 and higher.

While Scheduled Tasks can execute any Windows batch file or executable, Scheduled Jobs are specific to PowerShell and are used to generate and create background jobs that run once or on a specified schedule. Scheduled Jobs appear in the Windows Task Scheduler and can be managed with the scheduled task cmdlets of PowerShell. The only difference is that the scheduled jobs cmdlets cannot manage scheduled tasks.

These jobs are stored in the `Microsoft\Windows\PowerShell\ScheduledJobs` path of the Windows Task Scheduler. You can see and edit them through the management console in Windows after creation.

What's even greater about Scheduled Jobs in PowerShell is that you are not forced into creating a `.ps1` file for every new job you need to run. If you have a PowerCLI one-liner that provides all of the functionality you need, you can simply include it in a job creation cmdlet without ever needing to save it anywhere.

Getting ready

To being this recipe, you will need a PowerCLI window with an active connection to a vCenter server.

How to do it...

In order to schedule scripts using the native capabilities of PowerShell, perform the following steps:

1. If you are running PowerCLI on systems lower than Windows 8 or Windows Server 2012, there's a chance that you are running PowerShell 2.0 and you will need to upgrade in order to use this. To check, run `Get-PSVersion` to see which version is installed on your system. If less than version 3.0, upgrade before continuing this recipe.

2. Throw back a script you have already written, the script to find and remove snapshots over 30 days old from the *Removing targeted snapshots* recipe in *Chapter 5, Creating and Managing Snapshots*:

```
Get-Snapshot -VM * | Where {$_.Created -LT (Get-Date).
AddDays(-30)} | Remove-Snapshot -Confirm:$false
```

3. To schedule a new job, the first thing you need to think about is what triggers your job to run. To define a new trigger, you use the `New-JobTrigger` cmdlet:

```
$WeeklySundayAt6AM = New-JobTrigger -Weekly -At "6:00 AM"
-DaysOfWeek Sunday -WeeksInterval 1
```

4. Like scheduled tasks, there are some options that can be set for a scheduled job. These include whether to wake the system to run:

```
$Options = New-ScheduledJobOption -WakeToRun -StartIfIdle -
MultipleInstancePolicy Queue
```

5. Next, you will use the `Register-ScheduledJob` cmdlet. This cmdlet accepts a parameter named `ScriptBlock` and this is where you will specify the script that you have written. This method works best with one-liners, or scripts that execute in a single line of piped cmdlets. Since this is PowerCLI and not just PowerShell, you will need to add the VMware cmdlets and connect to vCenter at the beginning of the script block. You also need to specify the `-Trigger` and `-ScheduledJobOption` parameters that are defined in the previous two steps:

```
Register-ScheduledJob -Name "Cleanup 30 Day Snapshots"
-ScriptBlock { Add-PSSnapIn VMware.VimAutomation.Core; Connect-
VIServer servers; Get-Snapshot -VM * | Where {$_.Created -LT (Get-
Date).AddDays(-30)} | Remove-Snapshot -Confirm:$false} -Trigger
$WeeklySundayAt6AM
-ScheduledJobOption $Options
```

6. You are not restricted to only running a script block. If you have a routine in a `.ps1` file, you can easily run it from `ScheduledJob` also. For illustration, if you have a `.ps1` file stored in `c:\Scripts` named `30DaySnaps.ps1`, you can use the following cmdlet to register a job:

```
Register-ScheduledJob -Name "Cleanup 30 Day Snapshots"
-FilePath c:\Scripts\30DaySnaps.ps1 -Trigger $WeeklySundayAt6AM
-ScheduledJobOption $Options
```

7. Even better, those modules you defined in the earlier chapters are also fair game for your scheduled jobs. In *Chapter 5, Creating and Managing Snapshots*, you wrote a module for `30DaySnaps`. Rather than scheduling the scheduled job and defining the PowerShell in the job, a more maintainable method can be to write the module and then call the function from the scheduled job. One other requirement is that Single Sign-On should be configured so that the `Connect-VIServer` works correctly in the script:

```
Register-ScheduledJob -Name "Cleanup 30 Day Snapshots"
-ScriptBlock {Add-PSSnapIn VMware.VimAutomation.Core; Connect-
ViServer server; Import-Module 30DaySnaps; Remove-30DaySnaps -VM
*} -Trigger $WeeklySundayAt6AM -ScheduledJobOption $Options
```

How it works...

This recipe leverages the scheduled jobs framework developed specifically for running PowerShell as scheduled tasks. It doesn't require you to configure all of the extra settings as you have seen in previous examples of scheduled tasks. These are PowerShell native cmdlets that know how to implement PowerShell on a schedule.

One thing to keep in mind is that these jobs will begin with a normal PowerShell session—one that knows nothing about PowerCLI, by default. You will need to include `Add-PSSnapIn VMware.VimAutomation.Core` in each script block or the `.ps1` file that you use with a scheduled job.

There's more...

There is a full library of cmdlets to implement and maintain scheduled jobs. You have `Set-ScheduleJob` that allows you to change the settings of a registered scheduled job on a Windows system.

You can disable and enable scheduled jobs using the `Disable-ScheduledJob` and `Enable-Scheduled` job cmdlets. This allows you to pause the execution of a job during maintenance, or for other reasons, without needing to remove and resetup the job. This is especially helpful since the script blocks are inside the job and not saved in a separate `.ps1` file.

You can also configure remote scheduled jobs on other systems using the `Invoke-Command` PowerShell cmdlet. This concept is shown in examples on Microsoft TechNet in the documentation for the `Register-ScheduledJob` cmdlet.

In addition to scheduling new jobs, you can remove jobs using the `Unregister-ScheduledJob` cmdlet. This cmdlet requires one of three identifying properties to unschedule a job. You can pass `-Name` with a string, `-ID` with the number identifying the job, or an object reference to the scheduled job with `-InputObject`. You can combine the `Get-ScheduledJob` cmdlet to find and pass the object by pipeline.

See also

▶ To read more about Microsoft TechNet: **PSScheduledJob Cmdlets**, refer to
   `http://technet.microsoft.com/en-us/library/hh849778.aspx`

8

Performing ESXCLI and in-guest Commands from PowerCLI

In this chapter, you will cover the following topics:

- ▶ Retrieving the ESXCLI object in PowerCLI
- ▶ Using the ESXCLI vm namespace to kill a misbehaving VM
- ▶ Performing ESXi ping with an ESXCLI object
- ▶ Configuring custom storage and path selection policies
- ▶ Configuring coredump settings for an ESXi host from PowerCLI
- ▶ Executing native commands inside the guest operating system from PowerCLI

Introduction

ESXCLI is the default command-line interface for configuring ESXi hosts. It was introduced with vSphere 5.0 In 2010 and unified command line configuration under a single command with a set of namespaces. ESXCLI can be run interactively on the console of a host (after enabling it in the **Troubleshooting Modes** menu), or it can be initiated remotely from a VMware Management Appliance, or from an installation of the VMware vSphere CLI.

PowerCLI can natively perform many of the same configuration tasks as ESXCLI, but there are times when ESXCLI is needed to make a configuration change that PowerCLI does not allow. One example is to set the storage path selection policies on a host, which cannot be done with PowerCLI.

Rather than having to switch between two command lines and syntax, PowerCLI also provides you with a way to create an ESXCLI object on a host and allows you to execute ESXCLI commands with this object. The ESXCLI object works like any other object in PowerCLI or PowerShell and there is very little new to learn, except the structure of the ESXCLI object itself.

Even beyond ESXCLI, PowerCLI can be used to leverage scripts and commands inside guest operating systems. Because of the VMware Tools that run in virtual machines, PowerCLI can inject commands and routines to run and can receive the output of the commands to be used in PowerCLI.

Retrieving the ESXCLI object in PowerCLI

To begin leveraging ESXCLI from PowerCLI, you need to obtain an ESXCLI object and begin looking through the structure of the object. In particular, the methods attached to an ESXCLI object are powerful for performing configuration tasks on a host.

ESXCLI is limited in scope to an individual ESXi host. This means that it is not aware of vCenter and performing an operation on many hosts is going to require you to loop through a set of defined hosts to execute the same operation.

ESXCLI works differently than PowerCLI. Where PowerCLI is object-based and all of the data and methods for changing data are stored within objects, ESXCLI works on the concept of namespaces. There are 14 namespaces that comprise ESXCLI. If you run `esxcli` on the direct console of an ESXi host, the output is the list of namespaces with a description of each. The output is pictured in the following screenshot:

```
Available Namespaces:
  device       Device manager commands
  esxcli       Commands that operate on the esxcli system itself
               allowing users to get additional information.
  fcoe         VMware FCOE commands.
  graphics     VMware graphics commands.
  hardware     VMKernel hardware properties and commands for
               configuring hardware.
  iscsi        VMware iSCSI commands.
  network      Operations that pertain to the maintenance of
               networking on an ESX host. This includes a wide
               variety of commands to manipulate virtual networking
               components (vswitch, portgroup, etc) as well as local
               host IP, DNS and general host networking settings.
  sched        VMKernel system properties and commands for
               configuring scheduling related functionality.
  software     Manage the ESXi software image and packages
  storage      VMware storage commands.
  system       VMKernel system properties and commands for
               configuring properties of the kernel core system.
  vcloud       VMware vCloud Director
  vm           A small number of operations that allow a user to
               Control Virtual Machine operations.
  vsan         VMware VSAN commands.
```

On a clean installation of ESXi 5.5, there are 13 namespaces output. The `vcloud` namespace is added to an ESXi host once vCloud Director has been connected to the vCenter Server which manages the host. If you do not have vCloud Director running in your environment, the `vcloud` namespace is missing.

Getting ready

To begin this recipe, you will need a PowerCLI window and an active connection to a vCenter Server.

How to do it...

In order to retrieve the ESXCLI object in PowerCLI, perform the following steps:

1. The first step is to obtain your ESXCLI object. To do this, you use the `Get-EsxCli` cmdlet. If you perform a `Get-Help` cmdlet on `Get-EsxCli`, you will see that it requires a `-VMHost` parameter. This is not optional, and the easiest way is to obtain this from a `Get-VMHost` cmdlet and pipe it to `Get-EsxCli`. To begin, choose a single server, as shown in the following command line:

    `$esxcli = Get-EsxCli -VMHost esxsrv1.domain.local`

2. Once you have an object, the next step is to look at the namespaces from ESXCLI. To do this, enumerate the contents by calling the variable `$esxcli`. If you compare the output of this object to the output from running `$esxcli` on the same ESXi host, they match. The output is shown in the following screenshot:

```
VMware vSphere PowerCLI 5.8 Release 1
PowerCLI C:\Program Files\VMware\Infrastructure\vSphere PowerCLI> $esxcli

=====================
EsxCli: 192.168.0.241

   Elements:
   ---------
   device
   esxcli
   fcoe
   graphics
   hardware
   iscsi
   network
   sched
   software
   storage
   system
   vcloud
   vm
   vsan


PowerCLI C:\Program Files\VMware\Infrastructure\vSphere PowerCLI>
```

3. Next, explore the device namespace using dot notation. Type in the `$esxcli` variable name, a dot, and then the device to show the data within the device element. Look at the elements returned in the output from the object. Try it for the hardware and system elements too, as follows:

    `$esxcli.device`

    `$esxcli.hardware`

    `$esxcli.system`

4. Dig a bit deeper into the device element. The output from the previous step listed an alias element within the device. If you output this with the dot notation as `$esxcli.device.alias`, you will see two methods, `list()` and `get()`. Try running the `list()` method to see all of the device aliases:

    `$esxcli.device.alias.list()`

5. Next, do the same for the hardware element. Run `$esxcli.hardware` and you will see a `cpu` element listed. If you show the contents of the `cpu` element using `$esxcli.hardware.cpu`, you will see a method called `list()` available. Run the `list()` method in `hardware.cpu`. A screenshot of the output is shown just after the following command line:

    `$esxcli.hardware.cpu.list()`

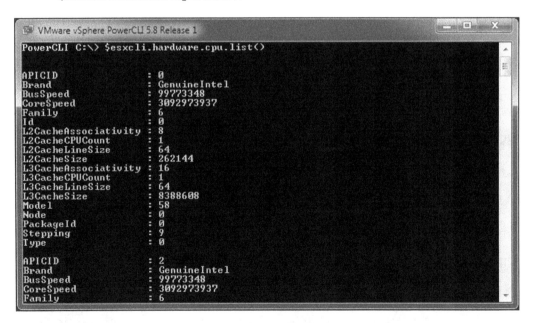

6. Like other PowerCLI objects, if you find data you want to use, you can assign it to a variable. In this case, you will use the `$cpuinfo` variable and assign the contents of the cpu list into it:

```
$cpuinfo = $esxcli.hardware.cpu.list()
```

7. Assigning data to a variable isn't the only thing that you want to extract and use it. You can also surround the command that returns the data with parenthesis. The parenthesis lets PowerCLI know how to execute the command inside it, and then, you can use dot notation behind the right parenthesis to access the properties returned. The following two commands return the same data. Using parenthesis is sometimes preferred, since it has shortcuts and uses a single line of PowerCLI code versus assigning it to a variable that requires two lines:

```
$cpuinfo.Brand
```

```
($esxcli.hardware.cpu.list()).Brand
```

How it works...

Working with an ESXCLI object should seem very similar to working with a View object in PowerCLI. Instead of calling the data within the object a property, this object calls the data in the object an element. In addition to elements, you also used a couple of methods. In the next few recipes, you will see that the two are very similar in which you have methods available that you can call to execute work within the ESXCLI object.

Since PowerCLI and ESXCLI work differently, PowerCLI adapts the namespace concept into elements within an object. The ESXCLI object represents each of the namespaces as an element and the subcommands of each namespace can be accessed using dot notation.

Fortunately, as you have already seen throughout this book, you can easily get a list of hosts from vCenter and loop through them in a `ForEach` loop. This is a great example of leveraging native ESXCLI capabilities and native PowerCLI capabilities to create a better solution than ESXCLI can offer on its own.

There's more...

There are many additional layers of data inside an ESXCLI object. For any `esxcli` command, you will find references in blogs or within the VMware documentation that can be adapted to run from PowerCLI. The advantage is that you can connect to a vCenter host with PowerCLI and create ESXCLI objects against many hosts to execute commands. With native `esxcli`, you must connect directly to a host each time, making scripting across many hosts very difficult.

Using the ESXCLI vm namespace to kill a misbehaving VM

After exploring the namespaces of ESXCLI, you can see that these are represented as elements in an ESXCLI object when used in PowerCLI. One of these namespaces or elements is the `vm` namespace. Within the `vm` namespace, you have commands that can list all of the running virtual machines on a host along with their process number. With this information, there is also a method in the `vm` namespace that allows you to kill a VM. If you are running PowerCLI to manage ESXi hosts, using the ESXCLI object is a quick way to access the virtual machine's `kill` command without enabling troubleshooting modes: the ESXi shell or SSH on an ESXi host. In this recipe, you will examine how to perform this operation.

Getting ready

To begin this recipe, you will need a PowerCLI window and an active connection to a vCenter Server.

 Caution! Do not execute this recipe on a VM unless you are sure you want to power it off. Running this on the wrong VM can cause data loss, because it immediately terminates the VM's process on the ESXi host and does not perform a graceful shutdown.

How to do it...

In order to kill a misbehaving VM using the ESXCLI `vm` namespace, perform the following steps:

1. The first step is to obtain an ESXCLI object, as you learned in the previous recipe. Remember that you must specify a single VMHost for your ESXCLI object, since ESXCLI is restricted to a single host at a time:

   ```
   $esxcli = Get-EsxCli -VMHost esxsrv1.domain.local
   ```

2. To find a VM, you will need to take a look at the `vm` namespace. To do this, use dot notation with your variable:

   ```
   $esxcli.vm
   ```

3. You will see that the `vm` namespace has one element and process. If you enumerate it, you will see two methods: `kill` and `list`:

   ```
   $esxcli.vm.process
   ```

4. While it's not necessary as often, sometimes, a VM stops responding, and for one reason or another you need to perform a `kill` operation on the VM. Storage problems can cause an issue such as these. In that case, you can list the process IDs for each VM on the host:

   `$esxcli.vm.process.list()`

5. If you check the output, the process ID is listed as `0` on all of your VMs, but you have a WorldID parameter defined. You can pass it to the `kill()` method in order to terminate the process. The `kill()` method requires a type and WorldID. The type is one of the three values: `soft`, `hard`, or `force`:

   `$esxcli.vm.process.kill("soft", 1000397005)`

Your WorldID will change for every VM each time it is started on a host. The example WorldID is for illustration purposes only, and should not be tried against your host.

6. If you list the running VMs again, notice that the VM you just killed is no longer listed. Even though the ESXCLI object is stored in a variable, this works in a slightly different way than normal with PowerShell variables, and the object is updated to reflect the current state after killing the VM.

How it works...

When you enumerated `$esxcli.vm.process`, PowerCLI returned a list of two methods that you can use. These equate to the same commands that can be used in ESXCLI. These commands in ESXCLI take switches to receive the additional information and make changes to the configuration, and in PowerCLI, these become methods with positional properties. Because these are positional, a value must be passed for each position, or at least to the last position so that you have an actual value to specify.

There's more...

There is much more that you can do with ESXCLI, and in the next couple of recipes, you will build some scripts to handle some common tasks where ESXCLI is required. ESXCLI has a full namespace devoted to network, but this is an area covered well by PowerCLI. Storage, on the other hand, is covered by PowerCLI, but the path selection policies and storage array type identification are not configurable from PowerCLI except through ESXCLI.

Performing ESXi ping with an ESXCLI object

While many of the networking configurations can be accomplished with native PowerCLI, one of the things in the networking space that might be useful from ESXCLI objects is the ability to perform ping testing from vmkernel interfaces on a host. By default, the `vmk0` interface is the default management interface of the host. In this recipe, you will use this interface and perform ping testing. Unlike a ping from a Windows or Linux system, the results of a PowerCLI ping will return in an object form just like any other cmdlet.

There are a couple of reasons to look at this particular use case. First, there are a lot of positional parameters with the ping. Second, some of the parameters aren't straightforward, and the in-line documentation doesn't explain much about parameters beyond their data types. However, since ESXCLI is the basis for these objects, all of the documentation for it also applies.

Getting ready

To begin this recipe, you will need a PowerCLI window and a connection to a vCenter Server. If we continue from the previous recipe, you can leverage the ESXCLI object and variable, and if not, you can obtain a new object in Step 1 of this recipe.

How to do it...

In order to perform ESXi ping testing from vmkernel interfaces on an ESXCLI host, perform the following steps:

1. The first step is to obtain your ESXCLI object. To do this, again, you use the `Get-EsxCli` cmdlet:

   ```
   $esxcli = Get-VMHost esxsrv1.domain.local | Get-EsxCli
   ```

2. Ping is found within the `network` namespace under the `diag` element. If you run `$esxcli.network.diag`, you will see a list of available methods, including ping.

3. Ping requires a number of positional parameters. The following is the output for the ping:

```
ping(long count, boolean debug, boolean df, string host, string
interface, string interval, boolean ipv4, boolean ipv6, string
netstack, string nexthop, long size, long ttl, string wait)
```

4. In all, there are 13 parameters, but luckily, you only need to specify up to the last parameter that matters to you. However, you must specify all of the parameters up to that point. In our example, you need to make sure that it is an IPv4 ping test and not an IPv6 test. You can ping an IP on the same network as your host for a count of 1 and for an interval of 1:

```
$pingresults = $esxcli.network.diag.ping(1, $false, $false,
"192.168.1.1", "vmk0", 1, $true, $false)
```

 One of the parameters specified is $false that is sent to the df parameter. This parameter is the Don't Fragment flag for the TCP traffic, but this isn't specifically outlined in the documentation on the command. To find the exact explanation, you can refer to the vSphere Documentation Center and the article about esxcli network. The link for this article is in the *See also* section of this recipe.

5. You will see that the results were stored in a variable. The variable contains two parameters: Summary and Trace. If you run a Get-Member cmdlet, you will see that both of these are of type Code Property. If you enumerate the value of the Summary property, you will see the success or failure of your ping:

```
$pingresults.Summary
```

6. If you have tested more than one count, you will have an array of results. You can enumerate the results one at a time using brackets and the position in the array. You can also loop through and display each with a ForEach loop:

```
$pingresults[0].Summary
```

How it works...

This recipe works based on a method from the ESXCLI object in the network namespace under the diag element. The ping performs a normal ping, but the method includes ways to force an IPv4 address ping only, an IPv6 address ping only, or multiple tests. You can specify the count of how many attempts and the intervals between each test.

There's more...

Even though you can use native PowerCLI cmdlets to configure networking properties, there is nothing to stop you from configuring networking using ESXCLI. The entire network namespace is available with commands to configure the network firewall in ESXi, the interfaces, their IP address or DNS search settings, and many more settings. All of these settings are given in detail in the vSphere Documentation Center for ESXCLI. Anything that can be done from a command line with ESXCLI can also be done in PowerCLI with the ESXCLI object.

See also

> ▶ For more information on **esxcli network commands,** visit the VMware vSphere Documentation Center at `http://pubs.vmware.com/vsphere-55/index.jsp?topic=%2Fcom.vmware.vcli.ref.doc%2Fesxcli_network.html`

Configuring custom storage and path selection policies

One of the more common uses for ESXCLI in PowerCLI is the configuration of **Storage Array Type Plug-in** (**SATP**) and **path selection policy** (**PSP**) settings. These configurations are particularly important from array to array. While ESXi ships with many default configurations, storage array vendors often have specific settings and claim rules that might need to be configured on each host. These settings can easily be set with ESXCLI, and since it works on the command line, it can work in PowerCLI with the ESXCLI object.

Getting ready

To begin this recipe, you will need a PowerCLI window and a connection to a vCenter Server. If we continue from the previous recipe, you can leverage the ESXCLI object and variable, and if not, you can obtain a new object in Step 1 of this recipe.

How to do it...

In order to configure custom storage and the path selection policies, perform the following steps:

1. The first step is to obtain your ESXCLI object. To do this, again, you use the `Get-EsxCli` cmdlet:

    ```
    $esxcli = Get-VMHost esxsrv1.domain.local | Get-EsxCli
    ```

2. To begin this recipe, first take a look at two custom SATP rules written in ESXCLI. You will need to adapt each into a PowerCLI routine in order to create the custom rule on each host:

```
esxcli storage nmp satp rule add -s "VMW_SATP_SYMM" -V "EMC" -M
"SYMMETRIX" -P "VMW_PSP_RR" -O "iops=1" -e "EMC Symmetrix (custom
rule)"


esxcli storage nmp satp rule add -s "VMW_SATP_ALUA" -P "VMW_PSP_
RR" -O iops=100 -c "tpgs_on" -V "3PARdata" -M "VV" -e "HP 3PAR
Custom iSCSI/FC/FCoE ALUA Rule"
```

3. The next step is to enumerate the `add` method in the ESXCLI to get a list of all the necessary positional parameters. These are outlined in the screenshot that follows the command:

```
$esxcli.storage.nmp.satp.rule
```

4. To adapt the EMC array's ESXCLI command to PowerCLI, you need to match the parameters to the correct order for PowerCLI:

```
$esxcli.storage.nmp.satp.rule.add($null, $null, "EMC Symmetrix
(custom rule)", $null, $null, $null, "SYMMETRIX", "iops=1", "VMW_
PSP_RR", $null, "VMW_SATP_SYMM", $null, $null, "EMC")
```

5. To adapt the 3PAR array's ESXCLI command to PowerCLI, you will need to do the same match in the correct order for the PowerCLI `add` method:

```
$esxcli.storage.nmp.satp.rule.add($true, "tpgs_on", " HP 3PAR
Custom iSCSI/FC/FCoE ALUA Rule", $null, $null, $null, "VV",
"iops=100", "VMW_PSP_RR", $null, "VMW_SATP_ALUA", $null, $null,
"3PARdata")
```

6. Sometimes, the SATP claims rule isn't the only change that needs to be made. It is common to reconfigure the existing LUNs to change their PSP for existing LUNs on a running system. This can be done without ESXCLI. It uses the `Get-ScsiLun` and `Set-ScsiLun` cmdlets. You can scope based on the `CanonicalName` parameter that matches your LUN identifiers and then force it to check your desired PSP:

```
Get-VMHost | Get-ScsiLun -CanonicalName "naa.600*" | Set-ScsiLun
-MultipathPolicy "roundrobin"
```

7. You might also want to scope based on a specific vendor name if you have multiple vendors or array types within your environment. Each array or vendor might have different best practices. In this case, use `Get-ScsiLun` and then use a `Where` statement. There is no parameter to specify a vendor on `Get-ScsiLun`. For this step, change the vendor name to `3pardata` unless you are running a 3PAR array:

```
Get-VMHost | Get-ScsiLun | Where {$_.Vendor -eq "3pardata"} | Set-
ScsiLun -MultipathPolicy "roundrobin"
```

8. Setting the vendor from ESXCLI is also possible, but it is going to involve the `list` method and a `ForEach` loop to set the PSP:

```
$luns = $esxcli.storage.nmp.device.list()

ForEach ($lun in $luns) {

$esxcli.storage.nmp.device.set($null, $lun.Device, "VMW_PSP_RR")

} <# End ForEach #>
```

9. In addition to setting the PSP, you can also set additional device configurations on the item. To do so, you can tap into `$esxcli.storage.nmp.psp.roundrobin.deviceconfig.set()`. Just add this line into the `ForEach` loop created in the previous step and both of the settings can be completed at once:

```
$esxcli.storage.nmp.psp.roundrobin.deviceconfig.set(0, 1, $lun.
device, [long]100, "iops", $false)
```

How it works...

The ESXCLI storage namespace commands configure the storage subsystem in ESXi. The PowerCLI equivalent in an ESXCLI object performs the exact same thing, and in this recipe, you work with the SATP claim rules and the PSP defaults and settings. This allows you to create default rules for the configuration of new LUNs and any pre-existing LUNs on a system following a reboot.

In this recipe, it was necessary to set all of the available parameters for the SATP claim rules since we needed to set the Vendor and it was the last value. Anything that was not specifically specified in the original ESXCLI command received a `$null` value in the PowerCLI command string, and the result is the same whether you use `esxcli` natively or through an object in PowerCLI.

As you saw with the path selection policies, it was necessary to go into two different areas: the `nmp.device` and the `nmp.psp` namespaces, for configuring different settings. First, you need to make sure that the device is using the round robin path selection policy and then you can access the `roundrobin` settings to configure it further.

There's more...

Similar to the network commands in the network namespace, the storage namespace has a lot of available methods that can be invoked by PowerCLI. These methods are of a wide range and can allow you to configure snapshot resignaturing (useful in disaster recovery or restoration scenarios), reset storage subsystems, list the devices connected to a server and their settings, and as this recipe illustrated, set a number of configuration values.

See also

▶ For more information on **esxcli storage available** refer to the VMware vSphere 5.5 Documentation Center at `http://pubs.vmware.com/vsphere-55/index.jsp#com.vmware.vcli.ref.doc/esxcli_storage.html`

Configuring coredump settings for an ESXi host from PowerCLI

While not as frequent as a Windows Blue Screen of Death, VMware's similar purple diagnostic screen is something you will run into from time to time due to a hardware problem, a driver issue, or a firmware issue on an ESXi host. In these cases, collecting diagnostic information from a crash is very important to diagnose and solve the problem. At the same time, it has become very common to boot your ESXi host from a USB or SD media rather than from traditional hard disks.

If you use a USB or SD media to boot, you need to send coredumps to a persistent location. You have several options. VMware ships a network-based coredump collector that installs with vCenter. You can specify a network location to send coredumps to. You can also save them to a file or partition on the server. Coredumps to file was added in vSphere 5.5. Setting these settings is extremely easy to do with ESXCLI, but there aren't native PowerCLI cmdlets to handle this. So, this is another great use for ESXCLI in PowerCLI.

In this recipe, you will configure a host to send both a network coredump and write it to a file.

Getting ready

To begin this recipe, you will need a PowerCLI window and an active connection to a vCenter Server.

How to do it...

In order to configure the coredump settings for an ESXi host to send both a network coredump and write it to a file from PowerCLI, perform the following steps:

1. The first step to set the coredump configuration is to obtain an ESXCLI object:

   ```
   $esxcli = Get-VMHost esxsrv1.domain.local | Get-EsxCli
   ```

2. Once you have the object, you will need to work in the system namespace again. Under system, you have the coredump element, and if you enumerate it you will see the three elements: `file`, `network`, and `partition`:

   ```
   $esxcli.system.coredump
   ```

3. Inside each of these namespaces/elements, there is a `get()` or `list()` method that allows you to see the current settings configured. For the network element, `get()` returns the settings. For the file element, `get()` returns the single active file. For file, there is also a `list()` method that returns all of the configured files, including inactive ones:

   ```
   $esxcli.system.coredump.network.get()
   ```
   ```
   $esxcli.system.coredump.file.list()
   ```

4. To set up a network coredump location, you will use the network element. There is a `get()` and a `set()` method in this element. To set this, the syntax is as follows:

   ```
   boolean set(boolean enable, string interfacename, string
   serveripv4, long serverport)
   ```

5. To assemble a command, you will need an interface name—a vmkernel port—that can talk to the network location that will receive your remote dump file. You will also need a destination IP address and a port number. Lastly, you need to enable it:

   ```
   $esxcli.system.coredump.network.set($null, "vmk0",
   "192.168.1.200", "6500")
   ```

6. If you notice, you did not set the enable positional parameter to $true in this command. If you try to change it to $true while setting the other parameters, the command will fail with an error that shows you can't combine the enable with other parameters. Instead, you have to call the enable parameter in a separate command after configuring the other settings:

   ```
   $esxcli.system.coredump.network.set($true)
   ```

7. If you do not have a network coredump collector, you can also set a file to save the coredump file into the collection. Unlike network coredump locations, you can have multiple file locations for coredumps, but only one can be active at a time. To add a file location, you use the `add()` method. The syntax for `add()` is as follows:

   ```
   boolean add(boolean auto, string datastore, boolean enable, string
   file, long size)
   ```

8. To add a new file, you need to specify the datastore name and the name of the file. These are the only two parameters required, but if you want to enable this new file, you can do it at the same time as you set the file (unlike the network settings):

```
$esxcli.system.coredump.file.add($null,"DATASTORE1", $true,
"CoreDumpFileName")
```

9. Since you have multiple files, you might need to change them. Using the `list()` method under the file element, you can get a path. With the path, you can use the `set()` method to change the active file. The syntax for the file `set()` method is shown as follows:

```
boolean set(boolean enable, string path, boolean smart, boolean
unconfigure)
```

10. To activate an existing file, the example command will consist of the path that includes the Datastore's UUID and the filename of the file setup, as shown in the following command:

```
$esxcli.system.coredump.file.set($true, "/vmfs/volumes/53cabdf3-
866f371f-c5bf-0017087d98f0/vmkdump/test.dumpfile")
```

How it works...

This set of commands works within the ESXCLI system namespace. The system namespace configures the system settings for ESXi hosts. The coredump settings are a set of commands related to the locations and destinations for where to send dump files after a crash of ESXi. The ESXCLI command allows files, a partition, or a network location. In this recipe, you created scripts to specify the network and file locations to save the coredump files into. The methods are different for the types of ways of setting up the locations and activating them. The same idiosyncrasies exist in ESXCLI directly, so these carry over to the methods that you use in PowerCLI.

There's more...

VMware has several knowledge base articles related to setting up coredump locations for ESXi hosts. Each of them has a lot of additional information about the best practices for setting up these settings. In particular, the network coredump article gives a lot of additional information beyond the scope of this section that is related to configuring and setting up the Dump Collector service that ships with vCenter Server. Configuring the Dump Collector service and configuring the host to send to the collector are both important settings in order to get this configuration to work properly.

It's also important to notice that you can mix and use both a file location and a network location to send coredumps. Since these files contain important diagnostic information to troubleshoot a crash, to have copies both on the filesystem and sent to the network collector is not a bad idea.

See also

▸ For more information on **ESXi Network Dump Collector in VMware vSphere 5.x** refer to VMware Knowledge Base available at `http://kb.vmware.com/kb/1032051`

▸ For more information on **Configuring ESXi coredump to file instead of partition** refer to the VMware Knowledge Base available at `http://kb.vmware.com/kb/2077516`

Executing native commands inside the guest operating system from PowerCLI

One of the most interesting cmdlets in PowerCLI is the `Invoke-VMScript` cmdlet. It allows you to execute scripts and commands inside the guest operating system if it is running VMware Tools. There are a lot of exceptions and requirements that need to be met in order for this cmdlet to work. First, forget about commands that require elevated privileges in Windows. If it prompts you with UAC, then it will most likely not work.

There are some prerequisites needed for `Invoke-VMScript` to work. First, you need to be running a 32-bit PowerCLI window, the VM guest must be running, and it needs to be running VMware Tools. You should be able to connect to port 902 on the ESXi host from the machine running PowerCLI.

Even with several requirements, this capability is handy. Some of the examples that are available on the Internet show starting and stopping application instances using `Invoke-VMScript` or setting the guest configuration values. In this recipe, you will put together a basic script that reports the proxy server from a Windows guest. You will also write a command that stops and then starts an Apache web server on an Ubuntu Linux server.

Unlike the previous examples with ESXCLI, `Invoke-VMScript` doesn't return its output as an object. It returns a simple string output back to the console.

Getting ready

To begin this recipe, you will need a PowerCLI window and an active connection to a vCenter Server. You will need a Windows guest running in a VM, and you will need the login credentials for an account inside the VM.

How to do it...

In order to execute the native commands inside the guest operating system from PowerCLI, perform the following steps:

1. The script assumes that you have an active connection to vCenter, but if not, run a `Connect-VIServer` cmdlet and connect to a vCenter Server.

2. The first step to perform an in-guest script is to assemble an `Invoke-VMScript` cmdlet. To see what is needed to create the script, use `Get-Help` on the cmdlet to view the parameters:

    Get-Help Invoke-VMScript

3. You will see multiple parameters devoted to credentials in the help menu. To connect to the VM, you will need the guest credentials passed in the command string. You can specify the credentials in the command line, but a more secure way might be to use the `Get-Credential` cmdlet to store credentials in a variable. You will receive a Windows credential box where you can specify your username and password for authentication to the guest operating system:

    $guestcreds = Get-Credential

     Keep in mind that the credentials are stored but not verified. The login won't occur until you run `Invoke-VMScript`, so you will not receive a prompt saying that the credentials are valid or invalid until later.

4. Looking at the parameters, the obviously needed parameters are `-ScriptText` for the command to run inside the guest, `-VM` to specify which VM to run the script inside of, and credentials to connect to the VM:

    Invoke-VMScript -ScriptText "echo Hello World" -VM "WinVM1"
    -GuestCredential $guestcreds

5. The `Invoke-VMScript` cmdlet can also run as single-line without an existing connection to an ESXi host or vCenter. In addition to the guest credentials, you would need to specify the details of the VMHost and its credentials to connect to the host:

    $hostcreds = Get-Credential

    Invoke-VMScript -ScriptText "echo Hello World" -VM "WinVM1"
    -GuestCredential $guestcreds -Server esxsrv1.domain.local
    -HostCredential $hostcreds

6. Beyond the simple Hello, World! example, you can perform real operations. Next, perform a `netsh` function in Windows to return the proxy settings of the system. The command is `netsh winhttp show proxy` for Windows 7 and Windows Server 2008 or later. Run the following command. You will see the screenshot following it, which shows you a sample of the output returned:

    ```
    Invoke-VMScript -ScriptText "netsh winhttp show proxy" -VM
    "WinVM1" -GuestCredential $guestcreds
    ```

7. Performing an in-guest operation on a Linux server is just as simple. Collect a set of credentials to connect to the Linux server and then pass a command to be executed. In this case, you can pass the service command to restart the `apache2` web server on the guest:

    ```
    $linuxcreds = Get-Credential
    Invoke-VMScript -ScriptText "service apache2 stop" -VM Ubuntu14
    -GuestCredential $linuxcreds
    Invoke-VMScript -ScriptText "service apache2 start" -VM Ubuntu14
    -GuestCredential $linuxcreds
    ```

    The output for the preceding command line is given in the following screenshot:

How it works...

The `Invoke-VMScript` cmdlet works using the VMware Tools to pass a script into the guest operating system and execute it. It requires the appropriate guest OS credentials to execute the code, so it adheres to the standard Windows security practices and will not allow an unprivileged user to inject or do anything more than they are allowed in the guest VM. However, from an administrator's standpoint, this allows a VMware administrator to script and handle some common guest configurations directly from PowerCLI. The net results can be more uniform deployments and less human errors possible by creating repeatable scripts for common tasks.

There are several limitations to the functionality. Its ability to perform configuration tasks can be hampered by the Windows `User Account Control` (UAC) functionality that tries to reduce the runspace privileges of administrative accounts within the operating system. On the Linux front, its capability isn't universally available to all flavors of Linux.

There's more...

There are several great blog posts about the `Invoke-VMScript` capability from some great PowerCLI community members. One post is from Luc Dekens and includes a full script for evaluating whether or not the requirements of `Invoke-VMScript` are met. This script is a great starting point that will help you verify that everything needed is set up for your attempts to run `Invoke-VMScript`.

See also

▶   For more information, refer to Luc Dekens's blog on **Will Invoke-VMScript work?**,
    available at `http://www.lucd.info/2012/01/01/will-invoke-vmscript-work/`

9
Managing DRS and Affinity Groups using PowerCLI

In this chapter, you will cover the following topics:

- ▶ Applying recommendations for partially automated DRS clusters
- ▶ Setting a cluster into maintenance mode with PowerCLI
- ▶ Using native DRS rule cmdlets to manage KeepTogether and Separate rules
- ▶ Learning the MoRef way of identifying objects
- ▶ Creating a DRS group for virtual machines
- ▶ Listing the members of a DRS group
- ▶ Updating the members of a VM DRS group
- ▶ Creating a custom function to update members of a DRS group
- ▶ Creating rules to maintain memberships of DRS groups using a custom function
- ▶ Using Compare-Object to audit group memberships for differences

Introduction

The **Distributed Resource Scheduler** (**DRS**) feature of vSphere offers administrators a hands-off method of balancing workloads across compute clusters and datastores. VMware DRS can be run in a fully-automated mode or in a partially-automated mode for compute, where DRS makes recommendations based on standard deviations and other mathematical computations for balancing workloads in a vSphere cluster. Storage DRS is a storage-specific application of the functionality that uses factors such as latency and capacity to balance storage in a cluster of datastores.

As with anything that is fully-automated, there will always be exceptions that administrators have to account for. There can be virtual machines that should never coreside on the same host, such as domain controllers in the same forest. There can be virtual machines that should always be on the same host to improve the performance due to high amounts of network traffic between them. These sorts of exceptions are managed with DRS rules. vSphere clusters have KeepTogether and Separate rules for these cases that can be defined on clusters.

DRS also has the ability to set up affinity rules that are applied when virtual machines start or are running in a cluster. These rules can define a group of virtual machines and apply a specific requirement to them on a group of hosts. PowerCLI does not ship a cmdlet to manage DRS groups or affinity rules, but there are methods to programmatically configure the groups in some cases.

PowerCLI offers some basic cmdlets to handle DRS rules and their creation. In this chapter, you will take a look at the built-in functionality of DRS rules from PowerCLI, and then you will create some additional functions that can help you configure DRS rules, where PowerCLI doesn't deliver default cmdlets.

Applying recommendations for partially automated DRS clusters

PowerCLI offers you the ability to set a cluster into autopilot, but that is not a requirement to receive the benefits for DRS. DRS clusters can also be set into a partially automated mode that lets DRS make recommendations and allows the administrator to apply those recommendations. PowerCLI offers native cmdlets to retrieve a list of recommendations and allows them to be applied.

`Get-DRSRecommendations` and `Apply-DRSRecommendation` allow you to retrieve the current recommendations and then apply them. In vSphere 6, the `Apply-DRSRecommendation` cmdlet changes to `Invoke-DRSRecommendation`. The functionality is the same, but the verb changes. Like any other native cmdlet, you can take the results and scope the list of recommendations to apply. One important thing to note here is that the `Get-DRSRecommendation` and the `Apply-DRSRecommendation` cmdlets only apply to compute clusters and not to storage clusters.

Getting ready

To begin this recipe, you will need a PowerCLI window and an active connection to a vCenter Server. You will also need to ensure that the cluster you want to use is set in a partially automated DRS mode. If the cluster is in a fully automated mode, the system will automatically apply any recommendations, so there will be none to retrieve.

How to do it...

In order to apply recommendations for partially-automated DRS clusters, perform the following steps:

1. If you have DRS enabled inside your cluster, you can obtain a list of the waiting DRS recommendations by running the `Get-DRSRecommendation` cmdlet. It requires no parameters:

   ```
   Get-DRSRecommendation
   ```

2. Without any options, `Get-DRSRecommendation` retrieves recommendations for the entire vCenter Server across all clusters. If you specify the `-Cluster` parameter or if you pipe in a cluster object from a `Get-Cluster` cmdlet, both will retrieve the same recommendations because the same cluster object is passed to the cmdlet:

   ```
   Get-DRSRecommendation  -Cluster "BigCluster"

   Get-Cluster "BigCluster" | Get-DRSRecommendation
   ```

3. Recommendations have an associated priority with them. When you set the automation mode of a DRS cluster to `FullyAutomated`, you have the ability to set the level of automation. A priority is set on each recommendation. With the `Get-DRSRecommendation` cmdlet, you can also retrieve only specific priorities with the `-Priority` parameter:

   ```
   Get-DRSRecommendation -Cluster "BigCluster" -Priority 1,2,3
   ```

4. Of course, there might be times when you need to only select a particular priority, and you can do that with a `Where` cmdlet similar to the one you used with other cmdlets. But what can you use to identify specific recommendations? By default, the `Get-DRSRecommendation` cmdlet returns the `Priority` and `Recommendation` columns, but there is also a `Key` property that is a unique identifier for every recommendation made. This is easy to sort with:

   ```
   Get-DRSRecommendation | Select Priority, Recommendation, Key

   Get-DRSRecommendation | Where {$_.Key -eq "255"}
   ```

5. Lastly, you can retrieve and apply recommendations by piping the recommendation object into the `Apply-DRSRecommendation` cmdlet:

   ```
   Get-DRSRecommendation  | Where {$_.Key -eq "255"} | Apply-
   DRSRecommendation
   ```

6. The same is true if you want to apply all recommendations for priorities 1, 2, and 3:

   ```
   Get-DRSRecommendation -Cluster "BigCluster" -Priority 1,2,3 |
   Apply-DRSRecommendation
   ```

How it works...

The `Get-DRSRecommendation` cmdlet retrieves recommendations from the vSphere cluster and enumerates them in one or more `DRSRecommendation` objects. The objects contain the cluster and target, a virtual machine, along with the priority, recommendation, and the reason for the recommendation. The parameters for the cmdlet allow you to retrieve only recommendations for a particular cluster or for a particular priority level.

The priority level of a recommendation is based on a mathematical formula that includes the standard deviation of the load imbalance between the hosts in the clusters. The priority is based on a scale up to five with the lower numbers having more impact or importance. There are two common reasons for DRS recommendations. The first is balancing the average CPU loads on hosts and the second is satisfying affinity rules defined on the cluster. The second reason is controlled by user-defined rules that you will look at in the next recipe.

Once you return a list of recommendations and you have selected only the ones that you wish to apply, you can pass the object into the `Apply-DRSRecommendation` cmdlet that applies the recommended action. The `Apply-DRSRecommendation` cmdlet expects a `DRSRecommendation` object to be passed in, either through a variable or a pipeline. At this point, DRS applies the recommendations for the objects specified. Also note, in vSphere 6, the command changes from `Apply-DRSRecommendation` to `Invoke-DRSRecommendation`.

There's more...

Sometimes, DRS recommendations can't be applied. These cases are referred to as faults, and they have a reason associated with them and an explanation why they cannot be applied. Faults are displayed in a separate area in the vCenter Client and include the prevented action, as well as what type of problem was attempted to be solved. In many cases, the fault can be caused by a virtual machine being assigned to groups with opposite rules. In this case, one rule or the other must be applied, but if no clear winner can be determined, vSphere leaves the VM where it is currently running.

See also

▶ You can refer to the **Calculating the priority level of a VMware DRS migration recommendation in vSphere 4.0** topic of the VMware Knowledge Base, available at http://kb.vmware.com/kb/1007485

Setting a cluster into maintenance mode with PowerCLI

Maintenance is inevitable. vSphere's DRS features do a lot to help administrators when maintenance is required. vSphere has a maintenance mode feature for hosts in a cluster that prevents any virtual machines from being started or moved onto the host while maintenance is occurring. Hosts need to be placed into maintenance mode for changes and patching to occur.

For a cluster running in a fully automated mode, putting a host into maintenance mode will kick off a set of automated steps that will use vMotion to move all of the running virtual machines to other hosts. Using the vSphere Client, you can optionally move any powered off virtual machines to other hosts. In a partially automated mode, placing the host into maintenance mode will cause a number of DRS recommendations to be made that will need to be applied manually. You can use the previous recipe to apply these recommendations.

If you use PowerCLI to place the host into maintenance mode, you do not have the ability to move powered off virtual machines automatically like you do when using the vCenter Client. However, you can easily retrieve and move any powered off VMs using the `Get-VM` and `Move-VM` cmdlets.

Getting ready

To begin this recipe, you will need a PowerCLI window and an active connection to a vCenter Server with a cluster defined.

How to do it...

In order to set up a cluster into maintenance mode with PowerCLI, perform the following steps:

1. The first step to place a host into maintenance mode is to retrieve the correct host. The easiest way to do this and confirm that you have only selected the host or hosts you want is to use the `Get-VMHost` cmdlet. For instance, if you want to put the odd hosts, `esx1`, `esx3`, and `esx5`, into maintenance mode, you would need to retrieve them. You can combine a regular expression in the `Get-VMHost` cmdlet to retrieve just the three that you want. To do this, include square brackets and the three numbers you want them to be in the location you want them:

    ```
    Get-VMHost esx[135].domain.local
    ```

2. With the correct hosts retrieved, you can place them into maintenance mode. To do this, use the `Set-VMHost` cmdlet. The `Set-VMHost` cmdlet has a `-Status` parameter that accepts the `Maintenance` setting and will begin the process of putting a host into maintenance mode:

```
Get-VMHost esx[135].domain.local | Set-VMHost -State "Maintenance"
```

3. If the host is in a fully automated mode, all virtual machines running begin to move with vMotions caused by DRS recommendations. If the cluster is in a partially automated mode, your next step is to retrieve the recommendations and apply them. To do this, you will use the `Get-DRSRecommendation` cmdlet:

```
Get-DRSRecommendations -Cluster BigCluster | Apply-
DRSRecommendation
```

4. Next, you should check whether any powered off VMs are left on the host. In the event that the host has a problem with the patch or configuration change, you want to ensure that the VMs registered, but powered off, on this host have been relocated and can be started elsewhere. To do this, you will use the `Get-VM` cmdlet. Pipe the three hosts you want into the `Get-VM` cmdlet:

```
Get-VMHost esx[135].domain.local | Get-VM
```

5. With the list of VMs returned, the next step is to move them onto a different host. To do this, use the `Move-VM` cmdlet and specify any of the other hosts. Since the VMs are not running, it really makes no difference to which host you relocate these VMs:

```
Get-VMHost esx[135].domain.local | Get-VM | Move-VM -Destination
esx2.domain.local -Confirm:$false
```

6. At this point, the host is placed into maintenance mode and work can be completed on it—whether it's hardware upgrades, software patches, configuration changes, or hardware replacement.

There's more...

Bringing a host out of maintenance mode is even easier, since you don't need to worry about the offline virtual machines. To do this, you will use the same `Set-VMHost` cmdlet except that you will change the `-State` parameter to be `Connected`:

```
Get-VMHost esx[135].domain.local | Set-VMHost -State "Connected"
```

It would also be incredibly easy to create a function that does both of the maintenance mode invocation and moves all of the powered off virtual machines to other nodes. What's more, your function can automate maintenance mode for a manual DRS cluster by combining all of the steps of this recipe.

The one piece that you would need to alter is the `Move-VM` cmdlet to move the powered off VMs. Instead of specifying a host, you should use a method where you return a list of hosts in the cluster with the `Connected` status, run a `Where` statement. The name is not like the host you're putting in maintenance mode. You can use a `Select -First 1` cmdlet to find a different host to move the VM onto.

Using native DRS rule cmdlets to manage KeepTogether and Separate rules

PowerCLI has a number of built-in cmdlets to manage DRS rules. These rules allow you to keep virtual machines together and ensure that they stay separated within a vSphere cluster.

A simple example of two virtual machines that should never coreside on the same host would be domain controllers. You want to ensure that all of your domain controllers are not running on the single host in case a physical server fails, so you will set a Separate rule.

A simple example of two virtual machines that should always coreside on the same host would be an application server and its database that handles gigabits of traffic per second. You can achieve better throughput and performance of the database traffic if the two virtual machines are on the same host.

These are the two scenarios that are covered in this recipe.

Getting ready

To begin this recipe, you will need a PowerCLI window and an active connection to a vCenter Server with a vSphere cluster defined and running in vCenter.

How to do it...

Let's see how to use native DRS rule cmdlets to manage KeepTogether and Separate rules by performing the following series of steps:

1. To begin creating a DRS rule, you will need to use the `New-DRSRule` cmdlet. The cmdlet requires a name and a cluster to be specified. In addition to this, you need to specify the `-KeepTogether` parameter. It's Boolean and requires either `$true` or `$false` to be passed into it. A `-KeepTogether:$false` statement is used for a Separate rule. Last, you have to specify a list of virtual machine names:

   ```
   New-DRSRule -Name "Separate Active Directory" -Cluster BigCluster
   -KeepTogether:$false -VM DC01,DC02
   ```

2. To verify the rule you just created, use the `Get-DRSRule` cmdlet. The cmdlet requires the `-Cluster` parameter to be specified:

   ```
   Get-DRSRule -Cluster BigCluster
   ```

3. You will see that the `VMIds` property doesn't display a friendly VM name that is easy to understand. Instead, it contains the **Managed Object Reference** (**MoRef**) for each VM. You will take a look at the MoRef in detail in the next recipe. With the MoRef; however, you are able to look up the VM by creating a calculated expression. The `VMIds` property includes multiple VMs, so a `ForEach` loop is needed along with the `Get-VM` cmdlet and the `-ID` parameter:

```
Get-DRSRule -Cluster Bigcluster | Select Name, Enabled,
KeepTogether, @{N="VM";E={ ForEach ($VM in $_.VMIds) { (Get-VM -Id
$vm).Name } } }
```

4. To create a second rule and to keep two application servers together on the same host, you will use the same syntax as a Separate rule, but you pass `$true` to the `-KeepTogether` parameter:

```
New-DRSRule -Name "App One" -Cluster BigCluster
-KeepTogether:$true -VM App01,DB01
```

How it works...

Managing the KeepTogether and Separate rules in PowerCLI is very simple. The cmdlet is the same, and the only difference between the two types of rules is the `-KeepTogether` parameter's Boolean value. Beyond that value, the structure of these rules is the same. You set up a name, specify a cluster, and specify the virtual machines that are affected by this rule.

Once the rule is in place, DRS will attempt to enforce the rule on its next run, either by generating a recommendation that can be applied in a partially automated cluster or implementing the change if it can meet the conditions specified in a fully automated cluster. These rules generate recommendations that reference the affinity rules as the reason for the recommendation.

Learning the MoRef way of identifying objects

A MoRef is a unique identifier in the vSphere platform for every individual object. Even two objects of the same type and the same name have different MoRef identifiers. Because vSphere builds relationships based on MoRefs, you can easily rename objects, such as virtual machines or resource pools, without affecting memberships and associations.

While this recipe might seem like a bit of a tangent, it is a necessary one before moving on to building VM and Host Groups for VM to Host affinity rules. These rules do not have native PowerCLI cmdlets to manage them; therefore, you need to use object views and methods to create these configurations and those require MoRefs.

You have encountered MoRefs in other recipes throughout the book, such as the *Setting Storage DRS automation levels for individual virtual machines* recipe in *Chapter 4, Working with Datastores and Datastore Clusters*, and the *Getting alerts from vSphere environment* recipe in *Chapter 7, Creating Custom Reports and Notifications for vSphere*. In both of these cases, the managed object reference was used in the code to identify a single virtual machine.

Commonly, MoRefs are the identifiers that you must use when creating and working with configuration specifications and the `Get-View` cmdlet because the underlying relationships are specific or related to this unique identifier and not to a display name, which might change. This implicitly ties a single, specific object to the specification that is being defined.

Getting ready

To begin this recipe, you will need a PowerCLI window and an active connection to a vCenter Server.

How to do it...

Let's see how to use the MoRef to identify objects by performing the following series of steps:

1. There are several ways to retrieve the MoRef of a virtual machine. The easiest way is to use a `Get-VM` cmdlet and explore the `ExtensionData` property of the VM. The MoRef is one of the properties inside the `ExtensionData` property. Enumerate the data in `ExtensionData` using dot notation and enclosing the cmdlet in parenthesis:

   ```
   (Get-VM -Name TTYLinux1).ExtensionData
   ```

2. Inside the `ExtensionData` property, you will see the associated data of the virtual machine. Notice the `Parent`, `Datastore`, `Network`, and `ResourcePool` properties. They all include MoRef identifiers that point to the associated objects where the VM is stored and connected.

3. If you go a step further and enumerate the `MoREF` property, you will find that it is actually a two-part property that contains `Type` and `Value`. To do this, add `MoREF` to the `Get-VM` cmdlet in dot notation:

   ```
   (Get-VM -Name TTYLinux1).ExtensionData.MoREF
   ```

4. Go back to the datastore for a moment; if, for instance, you have a MoRef, you can easily look up the object by `-ID` as you did in the previous recipe. To do this for a datastore, use the `Get-Datastore` cmdlet with the `-ID` parameter:

   ```
   Get-Datastore -ID ((Get-VM -Name TTYLinux1).ExtensionData.
   Datastore)
   ```

5. While you have other ways to get the datastore of a VM, this method is also viable. The point is that there are multiple ways to retrieve or correlate objects in PowerCLI, depending on what type of object you have.

How it works...

MoRefs are unique identifiers assigned to objects at the time of their creation. Even if you create more objects with the same name, each one will be generated with its own unique identifier. Many of the rules and relationships in vSphere are built around the MoRef. The native PowerCLI cmdlets hides the MoRef from your view and instead uses friendly names to make things easier to use. However, behind any of the commands that are executed, the MoRef is being called to identify one specific VM at a time.

When you begin working with `Views` and `ExtensionData` properties, the MoRef becomes more visible since the object views are the underlying methods for performing the actual work against vSphere. You begin to see these more as you take a look at the relationships between objects in vSphere.

There's more...

The MoRef is used frequently when you look at the Web Services SDK on the vCenter Server located at `https://vcentername/mob`. This SDK web interface is a good resource for you to explore and drill down to objects within vSphere. You might be able to discover and find the data you are looking for more easily in the web interface which you can then translate back to the same object within PowerCLI.

To begin, let's take a look at the Web Services SDK. You need to browse to `https://vcentername/mob` and log in with an account that has vCenter privileges. Once inside, you can see the top-level `Properties` and `Methods` split into two sections. One of the properties is `Content`. If you click on Content, you see a lot of properties with the name `Manager`. These are all data managers that span across the entire vCenter installation.

One of the properties listed is `rootFolder`. This points to the `Datacenters` default root folder, where you created your datacenter object in the *Creating a virtual datacenter in vCenter* recipe in *Chapter 2, Configuring vCenter and Computing Clusters*. If you drill down into `rootFolder`, you will see the datacenter you defined in *Chapter 2, Configuring vCenter and Computing Clusters*, with the MoRef of that object in the `childEntity` property. If you drill into the `childEntity` value, you begin to see many more MoRef objects for alarms, datastores, and networks. At this point, it should become more clear that the same structure you see exposed in the vCenter Client is represented in the nested data objects through the Web Services SDK.

This tool can be a powerful aid when you are building and writing routines in PowerCLI but need a 'treasure map' to guide you to the data you are looking for.

Creating a DRS group for virtual machines

Using what you just learned about MoRefs, you can build your first virtual machine DRS group. DRS groups are accessible in the Cluster view object. In this recipe, you will define a new group in the Cluster view object. The new group will contain several virtual machines and you will also define a host group. Once these are defined, you can create a rule that ties a VM group to a Host group.

To begin building a DRS group, the first step is to take a look at the structure in vSphere where the DRS groups are defined. This is part of the cluster view object in PowerCLI. In this view, you can define the configuration specifications that define the DRS groups, both for virtual machines and hosts. The same method applies for both, differing only between the type of object that you define.

Getting ready

To begin this recipe, you will need a PowerCLI window and an active connection to a vCenter Server. You should also have a virtual machine DRS group created and virtual machines assigned to the group.

How to do it...

Let's create a DRS group for virtual machines by performing the following steps:

1. The first step is to get the cluster view for the cluster that we want to manage. In this case, the cluster is `BigCluster`. To get the view, first use the `Get-Cluster` cmdlet and then pipe that into `Get-View`:

    ```
    Get-Cluster "BigCluster" | Get-View
    ```

2. The next step is to examine the `Configuration` and `ConfigurationEx` properties in this object:

    ```
    (Get-Cluster "BigCluster" | Get-View).Configuration
    (Get-Cluster "BigCluster" | Get-View).ConfigurationEx
    ```

3. You will see that the `ConfigurationEx` property includes a `Group` property within it. If you output that object, you will find the list of existing groups:

    ```
    (Get-Cluster "BigCluster" | Get-View).ConfigurationEx.Group
    ```

4. If you have a brand new cluster with no groups created, you might not get any results. In that case, the results of a cluster with existing groups would return a list of groups. Running the output through the `Get-Member` or `GM` cmdlet would show you two object definitions from the results: the first is the `VMware.Vim.ClusterVmGroup` object, and the second is the `VMware.Vim.ClusterHostGroup` object. These are the types of objects that you will need to create.

5. You will need to create a new object for the VM group, but this object must be stored in the more generic `VMware.Vim.ClusterGroupSpec` type. The reason it must be of this type is because the `Group` property in `ConfigurationEx` is of this type, but contains the `VMware.Vim.ClusterVmGroup` objects. To create the new object, use the `New-Object` cmdlet:

   ```
   $group = New-Object VMware.Vim.ClusterGroupSpec
   ```

6. If you enumerate the `$group` variable now, you will see an `Info` and `Operation` property. By default, the `Operation` property takes the value of `add`. To use this, you will populate the `Info` property with a new `VMware.Vim.ClusterVmGroup` object:

   ```
   $group.Info = New-Object VMware.Vim.ClusterVmGroup
   ```

7. Next, you can name your new VM group. To do this, you can simply use dot notation to populate the `Name` property of the `Info` property in the `$group` variable:

   ```
   $group.Info.Name = "TTYLinux VMs"
   ```

8. Next, you should populate the VMs that belong to this group. You can use any cmdlet that you want to retrieve an object that contains the VMs that should be assigned to the group. In this case, you can match against the `TTYLinux` VMs using `Get-VM` with a pattern match. Once you have the object, loop through a `ForEach` loop to assign the virtual machines to the `Vm` property:

   ```
   $vms = Get-VM "TTYLinux*"
   ForEach ($vm in $vms) {
      $group.Info.Vm += $vm.ExtensionData.MoRef
   }
   ```

9. Now that you have the group built, the last step is to assign this back to the cluster view. To do this, you need to use the cluster view. It is easy to store the cluster view in a variable so that you can easily use it:

   ```
   $cluster = Get-Cluster "BigCluster" | Get-View
   ```

10. Inside the cluster view object, there is a method named `ReconfigureComputeResource_Task`. Although this doesn't directly explain that it is used to make a group change, this is the method that will do the work. However, this method calls for a specific object to be passed into it. The output from a `Get-Member` cmdlet on the cluster view shows you the expected syntax of `ReconfigureComputeResource_Task`:

    ```
    TypeName    : VMware.Vim.ClusterComputeResource

    Name        : ReconfigureComputeResource_Task

    MemberType  : Method

    Definition  : VMware.Vim.ManagedObjectReference
    ReconfigureComputeResource_Task(
    VMware.Vim.ComputeResourceConfigSpec spec, bool modify)
    ```

11. Your group definition stored in `$group` can actually be stored in this `ComputeResourceConfigSpec` object type, but we need to create one more object and nest the $group object inside it. Use `New-Object` to create a `VMware.Vim.ClusterConfigSpecEx` object:

    `$spec = New-Object VMware.Vim.ClusterConfigSpecEx`

    `$spec.GroupSpec = $group`

12. Finally, you need to call the `ReconfigureComputeResource_Task` method and pass the specification stored in `$spec` into it:

    `$cluster.ReconfigureComputeResource_Task($spec,$true)`

 At this point, you created a new VM group. To do the same for a host group, you only need to change the procedure in Step 6 to be of the type, `VMware.Vim.ClusterHostGroup`.

How it works...

This recipe works by leveraging the cluster view object in PowerCLI. This object has a lot of methods and data stored in it that relate to the definition of a compute cluster. In this case, you created a new `Group` object that is of a generic group object called `VMware.Vim.ClusterGroupSpec`. Inside this object, you define either a host or VM group object with the name and host or VM objects linked to it. The `Group` specification is then stored in an overall `VMware.Vim.ClusterConfigSpecEx` object and that object is used with a Reconfiguration method on the cluster view.

Although there are several nested levels of objects, together they form a specification that defines the group configuration. Using the `Get-Member` cmdlet is the best way to understand what types of each object are present in the nested data structure. When you see an example or a definition of the types using `Get-Member`, you can engineer a structure to define new data. However, you must leverage a method to initiate a change. Without this critical piece, a defined specification never gets moved into use. This is a critical piece of the definition. In the next recipe, you will also leverage the same data structure to list members of an existing group.

There's more...

Even though you've gone through the steps of creating a new group, you probably do not want to do these steps manually for more groups. There is a much easier way to handle this by creating functions. Rather than covering all of that in this recipe, you can read blogger Arnim van Lieshout's two excellent and elegant functions to handle creation of new DRS virtual machine groups and new DRS host groups.

This blog post is an excellent example of many other great resources and scripts available online; however, there is no replacement for understanding how a script works. This recipe teaches you the "how to" part of the equation, but van Lieshout's script is a preferred way to handle future creation. The same blog post also includes a script using the same techniques to create new VM to Host rules. This script can be used to combine created VMs and Host groups and defines rules with them.

See also

▸ For more information on Arnim van Lieshout's New DRS Group Functions you can refer to `http://www.van-lieshout.com/2011/06/drs-rules/`

Listing the members of a DRS group

Now that you know how to correlate a MoRef back to a VM name, and how to create a DRS group, the next step is to be able to audit and list the members of DRS groups. Getting the list of groups is fairly simple. In fact, you already attempted to do this in the previous recipe. However, the returned information will be in the form of MoRefs for VM or Hosts. You need to convert these so that you can see the membership.

Getting ready

To begin this recipe, you will need a PowerCLI window and an active connection to a vCenter Server. You should also have a virtual machine DRS group created and virtual machines assigned to the group.

How to do it...

In order to audit and list the members of DRS groups, perform the following steps:

1. To report the members of a DRS group, you can do this with the cluster view object. The first step is to get the cluster view again for the cluster you want to report using the `Get-View` cmdlet. The parameter you want to report is in `ConfigurationEx` property and in the `Group` subproperty:

   ```
   (Get-Cluster "BigCluster" | Get-View).ConfigurationEx.Group
   ```

2. You can add a `Select` statement to this in order to get the name of the group and change the MoRefs into VM or Host Names. To do this, you can create a calculated expression that runs a `ForEach` loop on the `Vm` property to look up each VM and return its name:

   ```
   (Get-Cluster "BigCluster" | Get-View).ConfigurationEx.Group |
   Select Name, @{N='VMs';E={ForEach ($vm in $_.Vm) { (Get-VM -Id
   $vm).Name }}}
   ```

3. You can also create a calculated expression to return the list of Hosts by name:

```
(Get-Cluster "BigCluster" | Get-View).ConfigurationEx.Group |
Select Name, @{N='VMs';E={ForEach ($vm in $_.Vm) { (Get-VM -Id
$vm).Name }}}, @{N='Hosts';E={ForEach ($vmhost in $_.Host) { (Get-
VMHost -Id $vmhost).Name }}}
```

4. The results returned will show whether a group is a VM or Host group based on the type of members that are returned.

How it works...

This recipe works based on the data stored in the cluster view object. Since this represents all of the DRS groups, both VM and Host groups, you can easily access them. The primary component of the recipe is to create calculated expressions to translate the MoRef objects into usable data that is more meaningful.

There's more...

You can leverage this code to create a function to get all of the virtual machines that are a member of the DRS group and return them like a `Get-Function` cmdlet would. To do this, you use the code you have generated in this recipe and wrap it as a process section, add the header, params, and name your function. This will allow you to repetitively call the function whenever it is in use in other scenarios.

One of the things that creating the function might allow you to do is to take a returned object and compare it against the desired set of items in another object. Since the `Compare-Object` cmdlet of PowerShell is pretty simple to use, as long as the two objects match in type and format, you should be able to compare and drop out differences. This becomes a powerful auditing tool for your DRS groups.

See also

▶ The *Using Compare-Object to audit group memberships for differences* recipe in Chapter 9, *Managing DRS and Affinity Groups using PowerCLI*

Updating the members of a VM DRS group

Returning back to the task of building a host and VM DRS groups, the MoRef in the previous recipe will be used extensively. The groups and their memberships will be created using object views, configuration specifications, and the MoRef of each VM that belongs to the group.

You might be wondering why should you create or maintain these types of groups from PowerCLI instead of through the GUI, if there aren't native cmdlets available to you. In PowerCLI, it is easy to assemble a group of objects in an object that meets the specific criteria. This is something you've been doing all throughout the book. You can take criteria such as "all VMs on datastores from storage array X" and easily search for them with the `Get-DatastoreCluster`, `Get-Datastore`, and `Get-VM` cmdlets. Once you have that list, you can update the DRS group to match. PowerCLI actually makes much more sense to update DRS groups through the native vCenter Client method.

The vCenter Client doesn't offer any of the enhanced search capabilities in the DRS group member pickers. You get a list of VMs and you can search only for members to be added by name. It is a manual process to verify that the VMs stored on storage array X are the same ones in the group.

It is impossible to use the vCenter Client to identify which VMs are missing from all groups. In scenarios such as Metro Storage Clusters, you might want every VM to have an affinity to one site or another and this is generally done with DRS groups and VM-Host affinity. However, if you can't verify or audit this in the vCenter Client, there is no way to know for sure.

In this recipe, you will create `host` and `vm` DRS groups using MoRefs.

Getting ready

To begin this recipe, you will need a PowerCLI window and an active connection to a vCenter Server. You might want to have a text editor to store parts of the code to turn them into a reusable function.

You will need to start with an existing DRS group. If you do not have an existing DRS group, you can create one in the vCenter Client, or by using the *Creating a DRS group for virtual machines* recipe earlier in this chapter. This recipe will follow the same basic structure as the creation process.

How to do it...

In order to update the members of a VM DRS group, perform the following steps:

1. In the *Creating a DRS Group for virtual machines* recipe, you used a view in order to make the changes to the cluster DRS groups. There is another method. In this recipe, you will use a cluster object instead of a cluster view object. To obtain this, use the `Get-Cluster` cmdlet. You will utilize the `ExtensionData` property to make the changes later in the recipe:

    ```
    $cluster = Get-Cluster "BigCluster"
    ```

There are many methods to solve most problems in PowerCLI. You can easily write the earlier recipe using a Cluster object instead of a Cluster View object also.

2. The next step is to create three new objects that you will use for the configuration change. From the *Creating a DRS group for virtual machines* recipe, you will need a configuration specification object, a virtual machine group object, and a generic group object:

```
$spec = New-Object VMware.Vim.ClusterConfigSpecEx

$group = New-Object VMware.Vim.ClusterGroupSpec

$group.Info = New-Object VMware.Vim.ClusterVmGroup
```

3. You need to set the operation of the group specification to be of the type `edit` to make the group membership changes:

```
$group.Operation = "edit"
```

4. The next step is to specify the name of the group to be edited. To do this, you set the Name property in the `ClusterVMGroup` object. You can do this using dot notation on the `$group` variable that contacts the group specification, since the `ClusterVMGroup` object is a property in the group specification:

```
$group.Info.Name = "TTYLinux VMs"
```

5. In the creation recipe, you assigned any VM that matched the `TTYLinux*` pattern into the group. In this update, you will only assign the `TTYLinux1` VM into the group to ensure that your code has worked. Use the `Get-VM` cmdlet to retrieve the VM object, then use a `ForEach` loop to populate these into the group object:

```
$vms = Get-VM TTYLinux1

ForEach ($vm in $vms) {

   $group.Info.VM += $vm.ExtensionData.MoREF

}
```

6. Next, you should assign the completed group object into the specification. The specification was stored earlier in a `$spec` variable, so we can use dot notation again to set the `GroupSpec` property:

```
$spec.GroupSpec += $group
```

7. Last, you need to actually make the change that is defined by the configuration you just outlined. To do this in the earlier recipe, you used the `ReconfigureComputeResource_Task` method. This method also exists in `ExtensionData` of the cluster object. To check this, use the `Get-Member` cmdlet on `ExtensionData`:

```
$cluster.ExtensionData | Get-Member
```

8. The syntax of this method is the same: it receives two parameters. The first parameter is the specification and the second is a Boolean to make the change or not. The `$spec` variable contains our completed specification and a `$true` Boolean value will make the change:

```
$cluster.ExtensionData.ReconfigureComputeResource_Task( $spec,
$true)
```

9. The last step is to confirm that the changes were made. You can do this using the vCenter Client, or by using the *Listing the members of a DRS group* recipe.

How it works...

The update works the exact same way as the creation script in terms of assembling a specification with multiple PowerShell objects nested in an overall specification. This data is passed into a method to make the change to the cluster configuration.

In this recipe, you used a slightly different method to achieve the same outcome. One reason this was included was to illustrate the possibility of multiple methods to achieve the same results. The cluster object's `ExtensionData` property contains many of the same abilities as a cluster view. Using one over the other is a personal preference since they achieve the same thing. Using the cluster object might be preferred since the view requires an additional cmdlet; however, both are completely correct.

Creating a custom function to update members of a DRS group

Updates to DRS group members will likely occur much more often than the initial creation of a DRS group in vSphere. Because of this, creating a function to easily manage and update the members of a DRS group is helpful. In this recipe, you will create two functions that are used to update VM DRS groups and Host DRS groups.

In writing the function, you will begin to take into consideration the type of data that can be passed into the function from your users. This is important to consider, because as you've seen, there can be multiple ways to signify a particular object in PowerCLI that all points back to the same virtual machine. You can have string data, you can pass in a virtual machine object, a MoRef, or a View. All of these are valid representations of a virtual machine. The same is true for other elements in vSphere, such as virtual hosts, datastores, and networking.

To begin writing the function, you will pull back some of the same code that you used in the previous chapters where you created functions, but you will add some additional sections to the functions you will create in this chapter. This function is completely based on the Arnim van Lieshout scripts referenced in the *Creating a DRS group for virtual machines* recipe.

Getting ready

To begin this recipe, you need to open the PowerShell ISE, which will make creating a function easier.

How to do it...

In order to create a custom function to update members of a DRS group, perform the following steps:

1. To begin, you will need to connect to a vCenter; however, first you need to add the PowerCLI Snapin using the `Add-PSSnapin` cmdlet. These cmdlets can be executed in the lower half of the window where PowerShell is executed. They do not need to be in the function file you are creating:

   ```
   Add-PSSnapin VMware.VimAutomation.Core

   Connect-VIServer vcenter.domain.local
   ```

2. Your next step is to set up a generic function in the ISE window. To begin, create the function definition in the script file area of the ISE:

   ```
   function Update-VmDrsGroup {
   ```

3. Next, include the notes block of the function. This is the area where the description of the function, its examples, and other information will be outlined. This information is used for online help with the `Get-Help` cmdlet:

   ```
   <#
     .SYNOPSIS
       Updates the VMs in an existing VM DRS group
     .DESCRIPTION
       Reconfigures an existing virtual machine DRS group to
       contain the group of virtual machines defined by the
       -VM property of the function.
     .PARAMETER Cluster
       The name of the cluster with the DRS group to update
     .PARAMETER Name
       The name of the VM DRS group to update
     .PARAMETER VM
       The list of virtual machines to place in the DRS group.
     .EXAMPLE
   ```

```
          Update-VmDrsGroup -Cluster BigCluster -Name "VM Group"
            -VM (Get-VM WinVM*)
          .EXAMPLE
          Update-VmDrsGroup -Cluster BigCluster -Name "VM Group"
            -VM WinVM1,DB1,App1
      #>
```

4. The next step is to define the parameters that are accepted by the function. You do this inside a `param()` block. You will have three parameters to define: `Cluster`, `Name`, and `VM`. Each parameter needs to be mandatory. The VM parameter should not accept the pipeline input. In this function, you can also add some additional functionalities using `HelpMessage`:

```
param (
    [parameter(valuefrompipeline = $false, mandatory = $true,
      HelpMessage = "Enter a list of VM names")]
        [PSObject] $VM,
    [parameter(mandatory = $true,
      HelpMessage = "Enter the name of a cluster")]
        [PSObject] $Cluster,
    [parameter(mandatory = $true,
      HelpMessage = "Enter the name of a VM DRS Group")]
        [String] $Name
)
```

5. The next step is to begin the processing of the function. To begin, you use the process script block with a curly brace:

```
process {
```

6. The first thing to do in the process block is to define the new objects that you will need in order to update the group. These are the same three objects that you created in the previous recipe:

```
$spec = New-Object VMware.Vim.ClusterConfigSpecEx
```

```
$group = New-Object VMware.Vim.ClusterGroupSpec
```

```
$group.Info = New-Object VMware.Vim.ClusterVmGroup
```

7. Once you have the objects created, you should begin to populate the specification. The first thing to populate is the operation type and that should be set to `edit`:

```
$group.Operation = "edit"
```

8. The next step is to populate the name of the group from the `$Name` parameter, which will be passed in by the user:

```
$group.Info.Name = $Name
```

9. Next, you should attend to the cluster assignment. Because the `-Cluster` parameter accepts multiple types of input, you need to standardize the input before using it in your script. You need a cluster object, which is of type ClusterImpl. If you receive anything else, you need to change it into a cluster object. To do this for the Cluster parameter, you can use a switch statement. Each variable has a type associated and includes a `gettype()` method to return the type for evaluation:

```
Switch ($Cluster.gettype().name ) {
  "String" { $Cluster = Get-Cluster -Name $Cluster }
  "ClusterImpl" { }
  "ClusterComputeResource" { $Cluster = Get-Cluster -Id $Cluster.
    MoRef }
  "ManagedObjectReference" { $Cluster = Get-Cluster -Id $Cluster }
```

10. If the type doesn't match the types you need to accept for the parameter, you should display an error back to the user. You can do this by defining a default in the switch statement that throws an error for any other type:

```
default { throw "The data specified for the -Cluster variable
does not match the expected types."}
```

11. Don't forget to close the switch statement:

```
}
```

12. Next, you should also do the data clean up for the `-VM` parameter. In this block, however, you have additional work. Not only do you need to possibly get the VM, but you also need to add these into your group objects. You can do both in the script block. This is the same method used in the Arnim van Lieshout scripts referenced in the *Creating a DRS group for virtual machines* recipe. If a string is passed in the `-VM` parameter, you need to use the string to get virtual machines, loop through them, and add those to `$group.Info.VM`. If you get either a VM object (from `Get-VM`) or a VM view (from `Get-View`) passed in, you can add the `MoREF` values in a single line using dot notation. Even though there are multiple objects in the `-VM` parameter, if you use dot notation, the MoRef for all of the objects is accessible and returned by referencing it once. In any case, the objects should be added to the `$group.Info.VM` property using the `+=` operator to preserve any existing data and add the current object to it:

```
if (($VM.gettype().name) -eq "String") { $VM = Get-VM -Name $VM }
ForEach ($item in $VM) {
  Switch ($item.gettype().name ) {
    "String" { Get-VM -Name $VM | %{ $group.Info.VM +=
      $_.ExtensionData.MoREF }}
    "VirtualMachineImpl" { $group.Info.VM +=
      $item.ExtensionData.MoREF }
    "VirtualMachine" { $group.Info.VM += $item.MoRef }
  }
}
```

13. Next, you should check to make sure that the VM list has data. If it is empty, you need to throw an error. To do this, use an `if` command and specify the name of the property. If it is populated, it returns `$true` and if not, `$false`:

```
if ($group.Info.VM) {
```

14. Next, assign the group object into the specification object:

```
$spec.GroupSpec += $group
```

15. Last, execute the method to update the group membership with the method on the cluster object:

```
$cluster.ExtensionData.ReconfigureComputeResource_Task( $spec,
$true)
```

16. Now, you need to handle a `$false` condition from the `if` statement with an `else` statement. Inside the `else` statement, you would want to throw the error and tell the user that the update failed and why it failed:

```
} else {
    throw "There are no Virtual Machines defined. Update failed."
}
```

17. Now, close the process block next with a curly brace:

```
}
```

18. You have one last closing curly brace, this time to close the function:

```
}
```

How it works...

The `Update-VmDrsGroup` function takes the individual cmdlets and specifications from the previous recipe and builds controls and structure around them to create the function. When writing a multi-purpose function, you have to plan for the different types of input that can be passed into the function and handle processing for those different types, if that is required.

You have already seen functions, but in this case, you added several new things. You added controls for multiple parameters, where there were three required or mandatory parameters. You added definition to accept one of the three through the pipeline. You added multiple examples and documentation for each parameter to the documentation block. You also checked for some errors, used the throw command to stop processing, and sent the output to the user to indicate why a failure occurred. All of these are things you need to consider when putting together code that others will use. While you might pass in a string of virtual machines every time, another user might choose to pass them in from a `Get-VM` cmdlet or even from View or an unknown object type. If you were only processing based on strings, it would cause a failure when attempting to run.

There's more...

The same function can easily be updated to perform the same task for Host DRS Groups on a cluster. The documentation would need to change in the documentation block, but the actual processing is very much the same. Instead of using `Get-VM`, you will use `Get-VMHost` and you'd have to check the `VMHostImpl` type (from `Get-VMHost`) and HostSytem type (from `Get-View`) instead of the VM object types. Include both in the script file and save it as a module for easy updates of both VM and Host DRS Groups. Combine these functions with the Arnim van Lieshout authored scripts and you have a power set of tools to manage DRS groups on your vSphere implementation.

See also

 ▸ For more information on Arnim van Lieshout's New DRS Group Functions you can refer to `http://www.van-lieshout.com/2011/06/drs-rules/`

 ▸ The *Creating a snapshot management module* recipe in *Chapter 5*, *Creating and Managing Snapshots*

Creating rules to maintain memberships of DRS groups using a custom function

In this recipe, you will create a list of rules that leverage the `Update-VmDrsGroup` function that you created in the previous recipe in order to keep assignments up to date in a very dynamic environment. These rules can be defined and then set as a scheduled job in PowerShell to ensure that the assignments are up to date at all times, even if administrators move a virtual machine to a different location.

Getting ready

To begin this recipe, you will need a PowerCLI window and the function that you created in the previous recipe. You will also need an active connection to a vCenter Server. All of the cmdlets will update groups on the cluster, `BigCluster`. You will need to create five groups in the vCenter Client to use these rules and updates. Place the vCenter VM in each of the groups, since you cannot create an empty group. If the groups contain different members after writing the rules, the update worked successfully. The groups to be created are `Production Servers`, `iSCSICluster VMs`, `Datastore1 VMs`, `WinVMs`, and `VM Group`, on the `BigCluster` vSphere cluster.

For this recipe, you will create several commands to update VM group memberships based on the following rules:

- Update a DRS VM Group called `Production Servers` to include the database VMs in the `Production` resource pool. This might be necessary to ensure licensing compliance if you only license a subset of hosts in a cluster to run a database application.

- Update a DRS VM group with all of the VMs in a Datastore Cluster named `iSCSICluster`. This is helpful if you have a datastore cluster from an array at one site in a Metro Storage Cluster configuration to set affinity of hosts in the same site to a local storage.

- Update a DRS VM group with the VMs from a datastore called `Datastore1`.

- Update a DRS VM group with a list of VMs that match the wildcard pattern of `WinVM*` for a group named `WinVMs`.

- Update a DRS VM group from a list of specific VMs that don't match any particular pattern. The VM list is `WinVM1`, `TTYLinux1`, and `DC02`.

How to do it...

In order to create rules to maintain memberships of DRS groups using a custom function, perform the following steps:

1. To create the first rule, you can use the `Get-ResourcePool` cmdlet to retrieve the desired pool and then pipe it to `Get-VM` and retrieve only the VMs in that resource pool. From there, you use this command as the `-VM` parameter on `Update-VmDrsGroup` and update the group specified by name on the cluster specified. In this case, the name is `Production Servers` and the cluster is `BigCluster`:

    ```
    Update-VmDrsGroup -Name "Production Servers" -Cluster "BigCluster"
      -VM (Get-ResourcePool "Production" | Get-VM )
    ```

2. For the next example, you can use the same method of retrieving the datastore cluster and then piping that into `Get-VM` to obtain the list of VMs in that cluster. Then, specify it as the `-VM` into the `Update-VmDrsGroup` function:

    ```
    Update-VmDrsGroup -Name "iSCSICluster VMs" -Cluster "BigCluster"
      -VM (Get-DatastoreCluster "iSCSICluster" | Get-VM)
    ```

3. You can also use the parameters of a `Get-VM` cmdlet to return the list. In this case, use the `-Datastore` parameter to return all VMs on that datastore and then pass the VMs into the `Update-VmDrsGroup` function:

    ```
    Update-VmDrsGroup -Name "Datastore1 VMs" -Cluster "BigCluster"
      -VM (Get-VM -Datastore "Datastore1")
    ```

4. Wildcards are certainly within scope too. In this case, you can specify a wildcard and pass it directly as a string on the `-VM` parameter into the `Update-VmDrsGroup` cmdlet and the function will transform into the necessary objects:

```
Update-VmDrsGroup -Name "WinVMs" -Cluster "BigCluster" -VM WinVM*
```

5. For a list of unalike, nonmatching virtual machines, you can still handle this in a single command. You can specify a list of virtual machines using comma separation in the `Update-VmDrsGroup` function in the `-VM` parameter:

```
Update-VmDrsGroup -Name "VM Group" -Cluster "BigCluster" -VM
WinVM1,TTYLinux1,DC02
```

How it works...

As you near the end of the book, this is one of those recipes that helps you tie up loose ends. This brings together native cmdlet operations combined with a pipeline and a custom function that you've written. Each of the rules can be scheduled to run at an interval to update the group membership to ensure that virtual machines are running where you want them to be running.

There's more...

When you go to the next level of automation, the virtual machine groups or host groups with DRS. You should also closely watch for DRS faults in the vCenter Client. Faults, if you remember, are any of the recommendations that cannot be applied due to some sort of conflict condition. In the event that your rules apply virtual machines into groups that make up competing DRS rules, you will get faults. In the event of faults, you will need to research through and find which rules are causing DRS not to apply its recommendations on a virtual machine. Faults do not impair the running of a virtual machine; however, the VM might not be running where you want it to be located in the vSphere cluster.

See also

▶ The *Using PowerShell native capabilities to schedule scripts* recipe in *Chapter 7, Creating Custom Reports and Notifications for vSphere*

Using Compare-Object to audit group memberships for differences

As mentioned in the *Listing the members of a DRS group* recipe, you can easily compare a rule to the existing configuration as an auditing procedure. In this recipe, you will retrieve the list of existing VMs in the group, and then compare that with a list of the VMs from the desired group. You will use the group created in the *Creating a DRS group for virtual machines* recipe named `TTYLinux VMs`.

To make the results interesting, compare against the virtual machines `WinVM1` and `TTYLinux1`. Only one of these should be a match, while the others should fall out as differences. Then, you can compare the existing groups against a wildcard pattern match of `TTYLinux*`.

Getting ready

To begin this recipe, you will need a PowerCLI window with an active connection to a vCenter established.

How to do it...

In order to use `Compare-Object` to audit group memberships for differences, perform the following steps:

1. The first thing is to obtain a cluster object and obtain all of the DRS groups and store them into a variable:

    ```
    (Get-Cluster "BigCluster").ExtensionData.ConfigurationEx.Group
    ```

2. Next, you want to get the results of just the group you want to check. To do this, you can use a `Where` command to scope the results:

    ```
    $group = (Get-Cluster "BigCluster").ExtensionData.ConfigurationEx.
    Group | Where {$_.Name -eq "VM Group"}
    ```

3. The next step is to get a list of the current virtual machines that should match the DRS group defined. Obtain this result using the `Get-VM` cmdlet:

    ```
    $vm = Get-VM WinVM1,TTYLinux1,DC02
    ```

4. The next step is to run a `Compare-Object` cmdlet between the two lists. The groups from the cluster object come in MoRef form, which is the easiest thing to compare against. Since you used `Get-VM`, you have the MoRef of the VMs stored in `ExtensionData.MoRef` on the virtual machine object:

    ```
    Compare-Object ($group.Vm) ($vm.ExtensionData.MoRef)
    ```

5. To clean up the results, create a calculated expression to display the name of the virtual machine instead of its MoRef. The results for the following command line are displayed in the screenshot following it:

    ```
    Compare-Object ($group.Vm) ($vm.ExtensionData.MoRef) | Select @
    {N="VM";E={(Get-VM -Id $_.InputObject).Name}}, SideIndicator
    ```

```
VMware vSphere PowerCLI 5.8 Release 1

PowerCLI C:\> $group = (Get-Cluster "BigCluster").ExtensionData.ConfigurationEx.
Group | Where ($_.Name -like "TTYLinux UMs")
PowerCLI C:\> $group


Um                  : {VirtualMachine-vm-123, VirtualMachine-vm-164,
                      VirtualMachine-vm-166}
LinkedView          :
Name                : TTYLinux UMs
UserCreated         :
DynamicType         :
DynamicProperty     :


PowerCLI C:\> Compare-Object ($group.Vm) ($vm.ExtensionData.MoRef) | Select @{N=
"UM";E={(Get-UM -Id $_.InputObject).Name}}, SideIndicator

UM                                         SideIndicator
--                                         -------------
WinUM1                                     =>
TTYLinux3                                  <=
TTYLinux2                                  <=


PowerCLI C:\>
```

6. On the left-hand side, you have the two VMs, TTYLinux2, and TTYLinux3 displayed because they match the TTYLinux* pattern that was used to create the group. On the right-hand side, you see the WinVM1 that does not match the TTYLinux* pattern. This is expected, but it illustrates how you'd see results in the event that something doesn't match. If the two objects match completely, you will see different results. To illustrate this, update the contents of the $vm variable with a new Get-VM run to match against TTYLinux*:

```
$vm = Get-VM TTYLinux*
```

7. Now, rerun the same compare statement from the earlier recipe and you will see a different result—no result actually. The following screenshot illustrates the expected behavior:

```
Compare-Object ($group.Vm) ($vm.ExtensionData.MoRef) | Select @
{N="VM";E={(Get-VM -Id $_.InputObject).Name}}, SideIndicator
```

```
VMware vSphere PowerCLI 5.8 Release 1

PowerCLI C:\> $vm = Get-UM TTYLinux*
PowerCLI C:\> Compare-Object ($group.Vm) ($vm.ExtensionData.MoRef) | Select @{N=
"UM";E={(Get-UM -Id $_.InputObject).Name}}, SideIndicator
PowerCLI C:\>
```

8. In the real world, if you had different results and wanted to update the list, you can easily do this using the $vm variable. This would be an easy check and update if different. To do this, you can automate this with an if statement. If they are different (which means that a result is returned), execute the Update-VmDrsGroup function, or else output a message saying they are not different:

```
If (Compare-Object ($group.Vm) ($vm.ExtensionData.MoRef) | Select
@{N="VM";E={(Get-VM -Id $_.InputObject).Name}}, SideIndicator) {

    Update-VmDrsGroup -Name "TTYLinux VMs" -Cluster
    "BigCluster" -VM $vm

} else {

    write-host "There were no differences."

}
```

How it works...

The Compare-Object cmdlet is a very useful command that lets you compare the data inside two different objects and output the differences. The SideIndicator column shows which side has the difference. This would allow you to see where changes have occurred in the vSphere environment based on changes to the rules you have defined. For the comparison to occur correctly, you need to ensure that the data on both of the objects is of the same type. While type conversion can occur with comparisons, for the best results we ensure that both types match. In this case, we used the MoRef since that was the data available without conversion from the DRS group. This same data was available in ExtensionData on the virtual machine object. This allowed for a quick and easy comparison.

The results were not in a format that were easy to read or act on, however, this was easily fixed by using a calculated expression to look up and display the friendly name of the virtual machine.

This also allows you to keep your groups up to date at any level of automation where you feel comfortable. If you didn't want the updates to occur automatically on scheduled intervals, you can run this audit report and then send the results by e-mail to an administrator who can then go and act upon the changes.

See also

▶ The *Sending output to CSV and HTML* and the *Using PowerShell native capabilities to schedule scripts* recipes in *Chapter 7, Creating Custom Reports and Notifications for vSphere*

10

Working with vCloud Director from PowerCLI

In this chapter, you will cover the following topics:

- ▸ Connecting to a vCloud environment
- ▸ Creating and managing organizations in vCloud
- ▸ Creating a new user in an organization using Views
- ▸ Creating an organization's virtual datacenter in vCloud Director
- ▸ Importing a vApp template into vCloud
- ▸ Configuring networking in a vCloud vApp
- ▸ Reassigning vApp VM network settings with PowerCLI
- ▸ Starting and stopping vApps and individual VMs in a vCloud

Introduction

When installing PowerCLI, the optional vCloud Director PowerCLI can be installed to let you manage vCloud environments. This addition allows customers to connect to and manage vCloud installations, either in a private installation using vCloud Director or in a hybrid or public cloud hosted on a vCloud Provider. For vCloud customers who do not handle backend administration on vCloud Director datacenters, VMware also offers the PowerCLI for Tenants distribution, which is a reduced set of cmdlets needed for an end-user. For the recipes in this chapter, you will need the full distribution of PowerCLI with vCloud Director PowerCLI. Connecting to vCloud is a little different than connecting to and managing vSphere with PowerCLI, but all of the concepts that you learned up to this point apply in a vCloud environment.

Think of vCloud as a superset of controls over the vSphere virtualization platform. vCloud extends the concepts of vSphere, but also adds the concept of providers and organizations. Multiple providers might have virtual datacenters on the same infrastructure and offer services, securely, to organizations from the same vCloud. Each provider might have multiple tenants and each tenant is segmented from each other. The vCloud Networking and Security component, also known as vShield, is required to keep the tenants separated and secure from one another.

In addition to controls and security, vCloud also adds network capabilities to encapsulate and extend the border of datacenters between physical locations. This is important for hybrid cloud deployments where parts of the tenant's Cloud are onsite and part of it is hosted with a provider. vCloud Director manages both the local and remote vCloud locations and bridges the two.

Last but not least, vCloud abstracts services and applications a bit more than vSphere by packaging them into a service catalog that is made available to customers through a self-service portal.

In this chapter, you will work with vCloud Director using the vCloud Director PowerCLI set of cmdlets included in the PowerCLI installation. If you did not install PowerCLI for Tenants when you installed PowerCLI on your workstation, you will need to rerun the installation. You need to select **Modify** from the Program Maintenance window and select **This feature will be installed on your local hard drive**, as shown in the following screenshot:

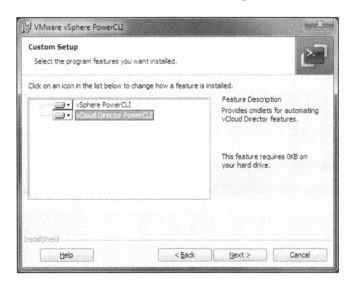

After the installation, the vCloud Director PowerCLI cmdlets will be available to you.

In order to complete the recipes in this chapter, you will need to have a vCloud environment to connect to. This can be an on-premise vCloud Director installation or a hosted vCloud environment with a service provider. This chapter assumes that you're acting as an administrator for at least your organization in vCloud.

Connecting to a vCloud environment

Before you can do any work in a vCloud environment, you need to connect to it. Connecting to vCloud Director or to a vCloud Hybrid Service Provider looks very similar to connecting to vCenter or an ESXi host.

Getting ready

The cmdlet to connect to vCloud is essentially the same as the `Connect-VIServer` cmdlet used to connect to vCenter or an individual ESXi host. Instead of the VI acronym, the cmdlets for vCloud Director use CI. So, the cmdlet to connect to vCloud is `Connect-CIServer`.

How to do it...

In order to connect to a vCloud environment, perform the following steps:

1. The first step is to connect using the `Connect-CIServer` cmdlet. Try using it just like a `Connect-VIServer` cmdlet:

    ```
    Connect-CiServer vcloud.lab.local
    ```

    If the certificate used to install vCloud Director is not trusted, you will be prompted with a message about an invalid certificate, as shown in the following screenshot:

2. You can choose to accept the certificate either one time or permanently. However, by default, it will deny access since the certificate is invalid.

3. You will be prompted to log in to the vCloud Director system with a Windows login prompt. If SSO is configured with vCloud Director, you can Single Sign-On without being prompted for the credentials.

4. Once logged in, you will return to the PowerCLI command prompt. To check the connection, you can enumerate a global variable called `$DefaultCiServers`:

```
$DefaultCiServers | Select *
```

How it works...

The first thing that you will see is the PowerCLI that displays a different message about the SSL certificate from the vCloud Director versus a vCenter with an untrusted certificate. If your certificate is not trusted for vCloud Director, the message, **It is recommended that you do not connect to any server with an invalid certificate**, is shown. The language and recommendation is slightly different, possibly because vCloud assets might be outside your datacenter and an invalid certificate can be a sign of a problem. Therefore, it is a very important factor when connecting to outside vCloud Director or vCloud assets hosted by third parties so that their certificates are valid.

If you are managing a vCloud Director with customers connecting to, you will need to ensure that you install a trusted certificate from a trusted certificate authority in order to avoid this issue for the customers. On internal deployments of vCloud Director, you can use an internal certificate authority that is trusted on your internal domain.

Beyond the handling and message of the certificate, if not trusted, the connection to vCloud Director is identical to the connection to vCenter Server, except for two parameters. The `Connect-CIServer` cmdlet allows you to pass a `-Org` parameter to connect to a specific organization. This is an important parameter that you will see in the following recipe. Using the `-Org` parameter ensures that you are connected to the desired organization's assets in vCloud. Since organization users cannot connect to the default vCloud, the `-Org` parameter is required for users in an organization. The other parameter is the missing `-AllLinked` parameter, since `linked-mode` is a vCenter function between multiple vCenter Servers.

There's more...

While in this recipe, you can connect as the super-user administrator for the entire infrastructure, the same cmdlet is used for an organizational administrator to connect and manage their infrastructure. In the next recipe, you will set up organizations and you will reuse this cmdlet to connect as the super-user administrator, and ensure that the security perimeters are defined between organizations.

Creating and managing organizations in vCloud

Organizations are a foundational concept of the vCloud environment. Organizations are used to separate tenants within a vCloud environment and can be interpreted in many ways depending on the use case for a vCloud deployment. For an internal vCloud provider, the organizations might represent departments that need to keep assets separated, such as in a university, governmental setting, or even a subsidiary of a company. For external vCloud providers, organizations might represent wholly separate businesses that have contracts with the provider.

Getting ready

To begin this recipe, you will need your vCloud environment running along with a PowerCLI window.

Before you can provision organizations in vCloud Director, you need to define a **Provider Virtual Data Center** (**Provider vDC**) in vCloud Director. If you have not performed this, you will need log in to the web interface and set this up.

How to do it...

In order to create and manage organizations in vCloud, perform the following steps:

1. The first step is to establish a connection with vCloud Director with the primary administrator account. To do this, use the `Connect-CiServer` cmdlet without an `-Org` parameter:

   ```
   Connect-CiServer vcloud.lab.local
   ```

2. Once you have the connection as the primary administrator, you will need to create a new organization. To do this, you will use the `New-Org` cmdlet. The `New-Org` cmdlet requires a `-Name` and `-FullName` parameter and has an optional `-Description` parameter. For this recipe, you will create several color organizations:

   ```
   New-Org -Name Red -FullName "Red Org"
   New-Org -Name Blue -FullName "Blue Org"
   New-Org -Name Orange -FullName "Orange Org"
   ```

3. Each of these organizations is now a perimeter or boundary where virtual assets can be deployed and managed.

How it works...

Creating Organizations within vCloud Director is very simple and takes a short, native `New-Org` cmdlet. This cmdlet doesn't take many options or configuration parameters, but it creates the basis on which everything else in vCloud Director is built. The organization is the base unit that is assigned to tenants in vCloud. Organizations form boundaries that contain the virtual infrastructure. Organizations can have different network definitions, different storage policies, and certainly separate accounts and permissions from one another. While many organizations in vCloud Director can share the same underlying vSphere infrastructure, the boundary is secure between organizations so that customers and their data never intermingle. It is possible, however, to share the network connections between two organizations using direct connections if two organizations need to communicate between one another.

As you move further in this chapter, you will explore additional, common administration and configuration tasks for vCloud Director. Each one goes a bit deeper into the capabilities while allowing you to ultimately build your first vApp and create a repeatable copy that can be deployed from a service catalog.

There's more...

To log in to any of the organizations defined in this recipe, you will need to create accounts. To do this, you use views since there is no native PowerCLI cmdlet to create a user in an organization. In the next recipe, you will perform this option and write a function that can be reused to handle this process.

Creating a new user in an organization using Views

In this recipe, since PowerCLI does not yet have a `New-CIUser` cmdlet, you will write a function to handle this process. The function requires to have the correct rights to create organizational users, but it will allow you to create users from PowerCLI without having to switch back to the web console. The code and function are based on a blog post from Alan Renouf, a product manager who focuses on automation frameworks and CLI at VMware.

Getting ready

To begin this recipe, open PowerShell ISE and set up a VMware PowerCLI environment by adding the PSSnapIn cmdlet, `VMware.VimAutomation.Core`.

How to do it...

In order to create a new user in an organization using Views, perform the following steps:

1. In PowerShell ISE, set up a basic function definition for `New-CIUser`:

```
Function New-CIUser {
```

2. Next, add your synopsis and help text to the function:

```
<#
.SYNOPSIS
   Creates a new user in a vCloud installation
.DESCRIPTION
   Creates a new user account in an organization of a vCloud
Director installation
.PARAMETER Name
   The username of the new user
.PARAMETER FullName
   The full name or display name of the user
.PARAMETER Password
   The password for the new user account
.PARAMETER Org
   The name of the organization to add the user into
.PARAMETER Role
   The role of the new user to be defined for the organization
.EXAMPLE
   New-CIUser -Name newuser -Password newpass -Org OrgName
-FullName "First Last" -Role "Organization Administrator"
#>
```

3. Next, create a parameter block and define the parameters needed to generate a new user account. The parameters should include a username, the display name, a password, the organization to add the user into, and the role of the user. The `Get CIUser` cmdlet also has an `Enabled` parameter where a user account can be disabled but it still exists. However, since you are creating an account, there is probably little chance that you need to create a disabled account:

```
Param (
   $Name,
   $Password,
   $FullName,
   $Org,
   $Role
)
```

4. Next, create a `Process` block in the function:

```
Process {
```

5. Now, you need to create a new user object to store the user information that will define the new account. The type of the new object is `VMware.VimAutomation.Cloud.Views.User`:

```
$orgAdminUser = New-Object VMware.VimAutomation.Cloud.Views.User
```

6. The next step is to populate the new object with the data supplied by the parameters on the function:

```
$orgAdminUser.Name = $Name
$orgAdminUser.FullName = $FullName
$orgAdminUser.Password = $Password
```

7. Since there is little need to create a disabled user account, you can statically define the `IsEnabled` property to `$true`:

```
$orgAdminUser.IsEnabled = $true
```

8. Next, you need to take the `$Role` parameter's data, which is most likely a string input and convert it into an object. You previously performed a procedure similar to this in the *Creating a custom function to update members of a DRS group* recipe in *Chapter 9, Managing DRS and Affinity Groups using PowerCLI*, using a `Switch` statement to identify the type of data passed into the function and to normalize this data to the object you need:

```
Switch ($Role.GetType().Name ) {
  "Reference" { $orgAdminUser.Role = $Role }
```

9. If you get a `String` input, you can try and match the string value against the defined objects and use these to pass in the `Role` object. To do this, first you need to obtain the roles defined in the vCloud Director. You can leverage the `$DefaultCIServers` global variable to find the role:

```
"String" {
  $vcloud = $DefaultCIServers[0].ExtensionData
```

10. Next, you can do a comparison using the `Where` statement to match the string provided against a defined role that is stored in `ExtensionData`:

```
$orgAdminRole = $vcloud.RoleReferences.RoleReference | Where {$_.Name -eq $Role}
```

11. With the role object capture, you can now assign it to the user object in the `Role` parameter:

```
$orgAdminUser.Role = $orgAdminRole
```

12. Close the string portion of the `Switch` statement with a right curly brace, and then close the `Switch` statement with a right curly brace:

    ```
        }
    }
    ```

13. The next step is to obtain the organization object's extension data. This is the data that contains additional details about the organization and includes a method called `CreateUser`:

    ```
    $OrgED = (Get-Org $Org).ExtensionData
    ```

14. Using the `CreateUser` method, you pass in the new user object and you are now able to create the user:

    ```
    $user = $OrgED.CreateUser($orgAdminUser)
    ```

15. Last, you can run a `Get-CIUser` cmdlet to check and ensure that the user account was created properly:

    ```
    Get-CIUser -Org $Org -Name $Name
    ```

16. Close the process block with a right curly brace:

    ```
    }
    ```

17. Close the function and it is ready for use:

    ```
    }
    ```

18. So, the next step is to highlight and run the object in PowerShell ISE. If there are no syntax errors, it should return you to a prompt with no output.

19. The next step is to try it with a new account. Create a user named, `orange1`, with a full name, `Orange Admin`, in the organization, `Orange`:

    ```
    New-CIUser -Name "orange1" -Password "newpass"  -FullName "Orange
    Admin" -Org "Orange" -Role "Organization Administrator"
    ```

20. The expected output will simply show you the output of `Get-CIUser`, the new user account once created. Any errors will be shown inline. If the user account is created properly, you can now try logging into the vCloud under the organizational administrator's account. When prompted, log in with the new user's credentials:

    ```
    Connect-CIServer vcloud.lab.local -Org Orange
    ```

How it works...

So, like any of the other recipes in the book created because of gaps in native PowerCLI cmdlets, this is another example of using Views and `ExtensionData` to do work. In this case, `ExtensionData` of the vCloud View allows you to create a user. It leverages the `CreateUser` method, but this method requires a user object to be passed into it. You can create and populate data into a new object of the expected type for the `CreateUser` method.

For the most part, the parameterized input is simply placed into the new user object, except for the role. The role is expected to be an object itself. Therefore, you have to check to see whether the user passed the role they wanted as a Reference object or if you need to find the corresponding Reference object to match the string that was passed in the command line.

Once the function is created, you can test it and then you can log into the Cloud with the organizational administrator's account in this recipe. Both of these are just tests to ensure that everything is working properly.

See also

▶ For more information on VMware PowerCLI Blog: **Automating creation of vCD Organizations, Users and Org vDCs** by Alan Renouf refer to `http://blogs.vmware.com/PowerCLI/2012/03/automating-creation-of-vcd-organizations-users-and-org-vdcs.html`

Creating an organization's virtual datacenter in vCloud Director

Within vCloud Directory's hierarchy, the Provider vDC is the top-level object, which is similar to the datacenter in vCenter Server. Inside a Provider VDC, you can house multiple organization virtual datacenters. In this recipe, you will create several Organization VDCs inside the provider VDC that is defined when you set up vCloud Director.

Each time that you create a new organization (tenant) on vCloud Director, you must give them a place to provision services. That is the purpose of the organizational VDC. However, organizations can have multiple Organization VDCs assigned to them, perhaps one onsite, and one at a third-party provider's site.

Unless you are handling multiple resellers on the same infrastructure, there is probably no need for multiple Provider VDCs, but there are always exceptions to any rule. One possible exception would be the infrastructure is managed by separate vCenter Servers. A Provider VDC is limited to a single vCenter Server. The point is that you have the ability to define multiple Provider VDCs, if needed, regardless of the reason.

In the previous recipe, you referenced a blog post from Alan Renouf. In the past, he has provided the code for a `New-OrgVDC` function. Since the post was created in 2012, vCloud Director PowerCLI has added a native PowerCLI cmdlet of the same name. Instead of relying on the custom function, you can now create virtual datacenters for organizations natively.

Getting ready

To begin this recipe, you will need a new PowerCLI window with an active connection to the vCloud Director under the primary administrator's account (not connected to a particular organization).

How to do it...

In order to create an organization's vDC in vCloud Director, perform the following steps:

1. As noted earlier, the cmdlet to create a new Organization VDC is `New-OrgVDC`. Run a `Get-Help` cmdlet on the cmdlet to get a view of all the parameters and the ways in which you can execute the cmdlet. There are actually three sets of execution parameters. The primary difference between the three parameters is the parameter to signify the Allocation Model.

2. To get a better understanding of the Allocation Models, you can move over to the vCloud Director web interface. Open the **Organizations** section under **Manage & Monitor** and select an organization. Click on **Manage VDCs** on the right-hand side of the page and then click on the **+** sign to create a new VDC. The Allocation Model screen is the second screen. The following screenshot is found under the **New Organization VDC** wizard and shows you the Allocation Models and descriptions for each:

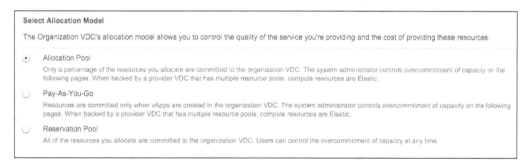

3. Beyond the Allocation Model and implications of each, the cmdlet parameters are the same between all the three sets of cmdlets. For the purpose of this recipe, you can use the `-AllocationModelAllocationPool` parameter:

4. To put together the cmdlet for the new Organization VDC, you will need to specify a name, the organization, the provider VDC, and any optional parameters. Both the organization and provider VDC have to be specified as objects, and not strings. The next step is to obtain these parameters:

```
$Org = Get-Org -Name "Orange"
$ProviderVDC = Get-ProviderVdc -Name "PrimaryVDC"
```

5. Next, you can assemble the New-OrgVDC cmdlet to create the new virtual datacenter. You will also need to specify the amount of CPU, RAM, and storage to be allocated to the new vDC:

```
New-OrgVDC -Name "Orange OnSite DC" -Org $Org -ProviderVdc
$ProviderVDC -AllocationModelAllocationPool
```

6. Now try the same cmdlet in the Blue Org, create it with the same parameters, but instead of passing variables in the command line, use nested cmdlets:

```
New-OrgVDC -Name "Blue OnSite DC" -Org (Get-Org "Blue")
-ProviderVdc (Get-ProvidedVdc -Name "PrimaryVDC")
-AllocationModeAllocationPool
```

7. If you forget to set anything, you need set it at the time of creation. You can always go back and perform the Get and Set operations on OrgVDC to change the settings:

```
Get-OrgVDC -Name "Orange Onsite DC" | Set-OrgVDC
-UseFastProvisioning $true
```

How it works...

This recipe is pretty simple, using a native cmdlet to provision the organization virtual datacenter. The only complication is that the cmdlet expects objects to be passed into the command line in order to identify the organization and the provider VDC. You can easily obtain these with additional cmdlets and pass that information into the command line.

In this recipe, you will use two different methods that work equally to create a vDC for an Organization. One uses inline cmdlets that utilizes parenthesis to pass the data objects into the main New-OrgVDC cmdlet.

Importing a vApp template into vCloud

The real utility of vCloud environments comes from the service catalogs and automated deployment of services. This orchestration makes it simpler for end users to go to a self-service portal and order a new instance of a service that is created and deployed on the backend infrastructure automatically.

The end user doesn't need to worry about which physical host to install the application, the database tier and the application tier, the network connections and IP pool of the application, and certainly not which types of storage should be utilized. All of these decisions should be made for the user based on the policy and definition of the vApp in vCloud.

The vApp is the building block of services in vCloud Director as well as vSphere. vApps package together the multiple pieces that create a full service or application. Many times, applications require multiple systems in order to operate, and the vApp concept helps them package together those disparate parts and wraps it with the configuration into a package.

The easiest way to get a vApp into vCloud Director is to import an existing virtual machine. The second way is to import an OVF or OVA format virtual appliance. The second method is easier to perform through the GUI, since PowerCLI doesn't provide you with a direct way to handle this through the command line. In PowerCLI, you have to import the vApp into vCenter Server and then pull it into vCloud Director using the first method. This recipe will focus on the first method.

Getting ready

To begin this recipe, you will also need a PowerCLI window with an active connection to vCloud Director and an active connection to vCenter. You will also need an existing virtual machine located in vCenter being managed by vCloud Director. This recipe is written to use the `TTYLinux1` virtual machine that was provisioned earlier in the book.

How to do it...

In order to import a vApp template into vCloud, perform the following steps:

1. The first step is to make sure that you are connected to both vCloud Director and vCenter Server for this recipe. Run both a `Connect-CIServer` and `Connect-VIServer` cmdlet and log in as prompted with credentials for each system:

   ```
   Connect-CIServer vcloud.lab.local
   ```

   ```
   Connect-VIServer vcenter.lab.local
   ```

2. You will need to connect to vCenter because you will use and pass a VM object from vCenter into an `Import-CIvApp` cmdlet for the vCloud Director to import the vApp from vCenter. `Import-CIvApp` requires the name of the VM, a new name for the vApp, and the organization to import the vApp into:

   ```
   Import-CIvApp -VM (Get-VM "TTYLinux1") -Name TTYLinux-vApp -OrgVDC
   (Get-OrgVDC -Name "Orange Onsite DC")
   ```

How it works...

The cmdlet for importing a vApp from an existing virtual machine is pretty simple. It requires just three parameters: `-Name`, `-VM`, and `-OrgVDC` to represent the new name of the VM, the VM object to import, and the Organization vDC objects respectively. Instead of using the `-OrgVDC` parameter, you can also use the `-Org` parameter if you only have a single Organization VDC for the organization. However, if you have more than one, you will need to specify the `-OrgVDC` parameter by its name because there would be no default values.

There's more...

You're not quite ready to start the new vApp you imported. There are still a few configuration changes required before you can fire up the new VM, including setting the network configuration. Even though the VM was configured and running on vSphere, the network configuration doesn't translate one to one into vCloud.

In the next section, you will investigate vCloud networking and begin to make the changes necessary to allow this VM to communicate with the network.

Configuring networking in a vCloud vApp

Before you can boot the vApps you just imported, you need to define networking for the vApp to connect to. While storage and storage policies are automatically pulled over from vSphere, the networking for vCloud requires a lot of additional configuration. This makes sense because connectivity to the services running on a vCloud have a lot of requirements, including private VLANs and Edge services to NAT addresses that are private to public or external facing networks. Some deployments might need to connect to preprovisioned VLANs directly to a virtual datacenter for client connectivity on the private-side of the cloud.

vCloud Networking can become an extremely deep subject and stray far beyond the scope of this book. However, there are some basic configuration things that are universal to vCloud Virtual Datacenters, and those are the topics this recipe will cover.

Again, the highest levels of configuration for vCloud Networking do not have native PowerCLI cmdlets to configure them. Instead, you should go into the vCloud Director GUI and preconfigure two portions of the networking configuration: an External Network, which will be a network connected to your live lab subnet, and a Network Pool, which will be a private network only in the vCloud that will be routed using vShield Edge to your live network.

Instead of focusing on the top-level provider network definitions, you will be faced with configuring the Organization VDC and vApp networks more often. The provider networks need to be connected and made available to the Organization VDC. Once this is done, the default vApp Network created with the `Import-CIVapp` cmdlet can be assigned and activated.

In this recipe, you will use Views in order to build a configuration and link the external network to the Organization VDC.

Getting ready

To begin this recipe, you will need a PowerCLI window and an active connection to vCloud Director.

How to do it...

In order to configure networking in a vCloud vApp, perform the following steps:

1. The first step to configure a newly imported vApp is to set up its network. At this point, your organization vDC does not have any network defined except for those attached from the provider VDC. To list the networks defined for the organization, use the `Get-OrgNetwork` cmdlet:

   ```
   Get-OrgNetwork -Org "Orange"
   ```

2. When you enumerate the networks, you should see the external network defined. The external network will be used to connect to vApp networks, so let's store the object representing this network in a variable. In this step, the recipe uses `Local` as the name for the external network, so change it according to your own environment:

   ```
   $ExtNet = Get-ExternalNetwork "Local"
   ```

3. The next step is to create a new object to define `OrgVdcNetwork`. This object is of type `VMware.VimAutomation.Cloud.Views.OrgVdcNetwork`:

   ```
   $OrgVdcNetwork = New-Object VMware.VimAutomation.Cloud.Views.OrgVdcNetwork
   ```

4. Name your new network in the `Configuration.Name` property:

   ```
   $orgVdcNetwork.Name = "Orange External Network"
   ```

5. Next, the `IsShared` property should be defined and set to `$false`:

   ```
   $orgVdcNetwork.IsShared = $false
   ```

6. Inside the `OrgVDCNetwork` object, you need to create a configuration object and store it in the `Configuration` property. This is where the options will be stored to define the network:

   ```
   $orgVdcNetwork.Configuration = New-Object VMware.VimAutomation.Cloud.Views.NetworkConfiguration
   ```

7. The first bit of configuration to perform is to set the upstream or parent network. This is the External Network that you want to connect to the Organization VDC. However, instead of referring to the External Network by an object, the `ParentNetwork` property actually refers to the network by a hyperlink. The hyperlink is stored in the `href` property of the External Network object:

```
$orgVdcNetwork.Configuration.ParentNetwork = $ExtNet.href
```

8. The next step is to set `FenceMode` to `bridged`. The three `FenceMode` options are: `isolated`, `bridged`, and `routed`. The `isolated` option passes no traffic, `bridged` passes native traffic from the connected network, and `routed` will configure and set up a vShield VM to route and perform other network functions between the networks:

```
$orgVdcNetwork.Configuration.FenceMode = 'bridged'
```

9. What is a network without IP addresses? Nonfunctional. Therefore, you need to define IP addresses, and to handle this, you will need to create another object and store it in the `Configuration.IpScopes` property:

```
$orgVdcNetwork.Configuration.IpScopes = New-Object VMware.
VimAutomation.Cloud.Views.IpScopes
```

10. Because this network is bridged, you can reuse the scopes defined in the parent network. Simply refer to the same data in the `$ExtNet` variable:

```
$orgVdcNetwork.Configuration.IpScopes.IpScope += $ExtNet.
ExtensionData.Configuration.IpScopes.IpScope
```

11. Last, but perhaps the most important is to create the network you have defined. To do this, you need to obtain the `OrgVdc` View:

```
$orgVdcView = (Get-OrgVdc -name "Orange Onsite DC").ExtensionData
```

12. Once you have the View, you can now use the `CreateNetwork()` method on the view to save the configuration you have defined:

```
$orgVdcView.CreateNetwork($orgVdcNetwork)
```

13. Now that the direct network connection has been extended into the Organization VDC, you can use it with the vApp you have imported. By default, the vApp brings a network map to its vSphere Standard Switch. You can map the Organization VDC network into the vApp. To do this, you use the `New-CIVAppNetwork` cmdlet, which requires the vApp and parent network to create a direct link:

```
New-CIVAppNetwork -VApp (Get-CIVapp -Name "TTYLinux-vApp") -Direct
-ParentOrgNetwork (Get-OrgNetwork -Name "Orange External Network")
```

How it works...

This recipe creates the network definitions in the organization vDC and inside a vApp, you need to connect the vApp that you imported and allow it to connect to the lab subnet. This is just one scenario, but this is a scenario not covered by the defined PowerCLI cmdlets. So, you need to turn to vCloud Views, similar to the Views you have leveraged for vSphere in the earlier chapters.

The Views work the same for vCloud infrastructure as vCenter controlled infrastructure. You create a new object and then populate data into the object, including additional objects linked to properties of the parent. Once a fully defined configuration is created, you can use a method from a View to execute and create the network. One difference to be noted in the steps is that other infrastructure objects, such as the reference to the parent network, are defined with hyperlinks and not with **Managed Object References** (**MoREF**) such as vSphere. This is one big difference using Views in vCloud versus vCenter.

Reassigning vApp VM network settings with PowerCLI

Now that you have the vApp imported and you have linked the outside network, both to the virtual datacenter for the organization and to the vApp, you can now assign it to the virtual machine inside the vApp. You can change any other vApp settings in order to make the application or service functional.

In the case of the imported vApp, you still have the default network imported with the vApp. To allow your TTYLinux-vApp to talk to the lab network, you will need to edit the virtual machine inside the vApp and map its network connection to the new vApp Network named Orange External Network that was defined in the previous recipe.

This time, PowerCLI has native cmdlets to perform the network reassignments. In this recipe, you will remove the imported network and reassign the VM to use the new network connection for the local lab network.

Getting ready

To begin this recipe, you will need a PowerCLI window with an active connection to vCloud Director.

How to do it...

In order to reassign vApp VM network settings with PowerCLI, perform the following steps:

1. The first step to reassign the vApp is to locate the current network connection, set it to none if it is not already set and remove any unneeded network assignments on the vApp. To check for network assignments on the vApp, use the simple `Get-CIvAppNetwork` cmdlet. To scope the results down to the single vApp you are looking for, first use a `Get-CIVApp` and pipe that result into `Get-CIVAppNetwork`:

   ```
   Get-CIVApp -Name "TTYLinux-vApp" | Get-CIVAppNetwork
   ```

   The output for the preceding command line is given in the following screenshot:

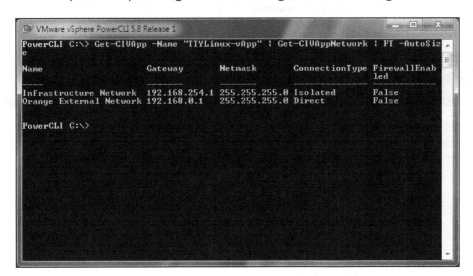

2. Your output should contain a couple of different networks. The **Infrastructure Network** parameter in the preceding screenshot was the imported network that came from vCenter, which is named the same as the Standard vSwitch assignment of the source VM. The connection type for this network is Isolated, which means that it will not communicate with the outside resources, so that won't help you connect your new vApp to the lab network so it should be removed. You will utilize the network you created in the previous step, which is the `Direct` connection type. Repeat the cmdlet and store the Isolated network in a `$IsolatedNet` variable and store the `Direct` connection in a `$DirectNet` variable. You will use both the networks in the next steps:

   ```
   $IsolatedNet = Get-CIVApp -Name "TTYLinux-vApp" | Get-CIVAppNetwork -ConnectionType Isolated
   ```

   ```
   $DirectNet = Get-CIVApp -Name "TTYLinux-vApp" | Get-CIVAppNetwork -ConnectionType Direct
   ```

3. The next step is to reassign the network adapter on the VM in the vApp. To do this, you will use the `Set-CINetworkAdapter` cmdlet. The cmdlet requires the network adapter to be passed into the cmdlet, and an easy way to do it is to pipe the output from `Get-CINetworkAdapter` to retrieve the adapter object. You can also use `Get-CIVApp` with the name of the vApp to change and `Get-CIVM` (along with the name of the VM if the vApp has multiple VMs) to scope down the list of the network adapters to the one you want to change. Since there is only one VM and one network adapter in the VM, the `Get-CIVM` and `Get-CINetworkAdapter` cmdlets require no parameters:

    ```
    Get-CIVApp -Name "TTYLinux-vApp" | Get-CIVM | Get-CINetworkAdapter
    | Set-CINetworkAdapter -VAppNetwork $DirectNet -Connected $true
    -IpAddressAllocationMode "Dhcp"
    ```

4. Since the Isolated network serves no purpose for connecting to the VM and it has now been replaced with the direct network connection on the VM. You can now remove it. Removing the network is simple with the native `Remove-CIVAppNetwork` cmdlet. Since you stored the isolation vApp Network in a variable earlier, you can simply pass the network object into the cmdlet with the `-VAppNetwork` parameter:

    ```
    Remove-CIVAppNetwork -vAppNetwork $IsolatedNet
    ```

An important thing to notice is that even if the vApp Network has the same name on each vApp, this object is specifically linked to `TTYLinux-vApp` and represents a specific network to be removed. Each object is unique and this can be verified with the `href` property on the object.

How it works...

This recipe was achieved with all native cmdlets. The trickiest part of the procedure is to retrieve the correct network adapter to be changed, but the specific network adapter can be easily found. You just need to know the hierarchy of where the network adapter is located. The network adapter needed is in the vApp named `TTYLinux-vApp` and in the VM, which is inside the vApp. String together the three native cmdlets and you will be able to retrieve that network adapter. Once you have done this, you can reassign the network with the `-vAppNetwork` parameter and also make any additional reconfigurations that you want.

There's more...

A vApp is a container. The container allows multiple components that make a fully functional app to coreside inside the vApp container. This means that multiple virtual machines, such as a web frontend VM and a processing VM, can be defined and a direct network link between them can be created. It also means that a third database VM can be created and linked to the processing VM on a separate private network, creating a three tier application infrastructure.

The web frontend can be the only one of the three that talks to the outside networks, and as an administrator, you might want to turn the firewall off for the web frontend. All of this can be defined in a vApp.

Once a vApp is defined, you can import it into the service catalog for easy and repeatable deployments of this app. All of the required infrastructure, including network services such as firewalls or load balancers, can be packaged so that when a new instance of the app is deployed, the fully functional package lays down all of the required virtual infrastructure to allow it to run in a self-contained environment.

For the task of designing multi-tier vApps, you'll likely use the vCloud Director web administration to configure all of the virtual machines and networks, but PowerCLI can easily take that package and deploy them in mass, as it is needed by the customers.

Starting and stopping vApps and individual VMs in a vCloud

Now that the vApp is configured, you are ready to start the vApp. Because of the vApp definition and packaging, even a vApp that contains only a single VM has multiple ways to control it. A single VM vApp can also contain security or network related virtual appliances that deploy and run along with the VM (in the case of routed vApp Networks).

vApps can define the boot order and can wait for one system to come online before starting the next using the boot order definitions. So, starting the vApp will bring systems up in an orderly way, where using vCloud VM cmdlets lets you control the VM boot regardless of the vApp boot orders. There are multiple cmdlets that can stop individual components of the vApp or the entire vApp entity.

In this recipe, you will start a vApp and then restart an individual VM in the vApp, just like you would if a single VM was misbehaving or if it might have locked up. You can also shut down an individual VM without stopping the vApp itself.

Getting ready

To begin this recipe, you will need a PowerCLI window with an active connection to vCloud Director.

How to do it...

In order to start and stop vApps and individual VMs in a vCloud, perform the following steps:

1. To start a vApp, you will use the `Start-CIVApp` cmdlet. It requires just a reference to the vApp. An easy way to do this is to perform a `Get-CIVApp` cmdlet and search for the vApp you want to start and pipe that to `Start-CIVApp`:

   ```
   Get-CIVApp -Name "TTYLinux-vApp" | Start-CIVApp
   ```

2. Starting a vApp can take an extended amount of time in vCloud, especially for the first time. This is because some vApps require additional network and security appliances to deploy and boot before the VMs themselves can start. The vApp packages can have other infrastructures to provision on first boot.

3. Once the vApp has started, you can check its status with the `Get-CIVApp` cmdlet.

4. If a single VM within the vApp misbehaves, you might want to quickly restart that single VM. To do this outside the vApp controls, PowerCLI has the `Restart-CIVM` cmdlet. This cmdlet needs the VM object to be passed into it, and since there can be many VMs with the same name in vCloud (remember, vApps have the same name for the VMs in each deployment), you should use `Get-CIVApp` and pipe that to `Get-CIVM` to get the specific VM you want to restart:

   ```
   Get-CIVApp -Name "TTYLinux-vApp" | Get-CIVM | Restart-CIVM
   ```

5. With vCloud, you cannot dynamically reassign a network on a VM since the network definition can include network and security appliances, network services, and IP address pools. So, you might need to shut down a VM that has misassigned a network, change the network, and start it again. PowerCLI has cmdlets specifically for stopping and starting a VM in a vApp. The same procedure for locating the VM with `Get-CIVApp` and `Get-CIVM` should be used to identify the correct VM:

   ```
   Get-CIVApp -Name "TTYLinux-vApp" | Get-CIVM | Stop-CIVM
   ```

6. Make your network or other VM configuration changes and run the `Start-CIVM` cmdlet:

   ```
   Get-CIVApp -Name "TTYLinux-vApp" | Get-CIVM | Start-CIVM
   ```

7. Finally, if you're ready to shut down and decommission a vApp, which might include many VMs and other configurations, you can use the `Stop-CIVApp` cmdlet to shut down the entire package. Then, you can use `Remove-CIVApp` to remove it from vCloud:

   ```
   Get-CIVApp -Name "TTYLinux-vApp" | Stop-CIVApp
   Get-CIVApp -Name "TTYLinux-vApp" | Remove-CIVApp
   ```

How it works...

The cmdlets from this recipe are very simple and straightforward. The native cmdlets for vApp and VM control are easy to use and control the vApp's operation in vCloud. The critical thing to keep in mind and the reason to call these out in the chapter is that with a vCloud deployment, there can be tens or hundreds of copies of a vApp on the infrastructure. This makes it critical, particularly for backend administrators, to ensure that they are working in the correct organization and the correct vApp to ensure that the correct vApp or VM is started, stopped, or restarted.

One way to assist you with this is to connect to an account with the least privileges required to administer what you're configuring. If you are working with a particular organization's datacenter, you will want to connect directly to that Organization VDC rather than to the primary vCloud account so that you only see the vApps deployed for that one company. This helps you avoid any unintended downtime or problems.

Setting up and Configuring vCloud Director

For the recipes of *Chapter 10, Working with vCloud Director from PowerCLI*, you will need a vCloud environment to connect to for testing the vCloud Director PowerCLI. There are several options for a test environment to connect to. In this section, there are three options that you can perform for establishing a vCloud Director environment in this book, and they are as follows:

- ▸ The hosted vCloud environment
- ▸ Deploying the vCloud Director environment from AutoLab
- ▸ Building your own vCloud Director implementation on your vSphere lab environment

The hosted vCloud environment

Perhaps the easiest method to test **vCloud Director PowerCLI** is a hosted vCloud environment. If you have an account with a hosting provider for a vCloud environment, and you have the administrative access to your organization, you can perform all of the recipes in this book.

Deploying the vCloud Director environment from AutoLab

If you don't have vCloud Director available in a lab environment, an automated option to set up an environment is the **AutoLab**, created by Alastair Cooke, and it is supported by many VMware community members. AutoLab is available from `http://www.labguides.com`, and it is a set of VM definitions for ESXi or for VMware Workstation and VMware Fusion to configure an entire, self-contained lab environment.

You will still need to obtain all of the software yourself; however, AutoLab fully automates the deployment of a functional lab environment. In additional to vSphere, AutoLab includes scripts to deploy vShield Manager and vCloud Director that were added by Damian Karlson. Deploying AutoLab will take several hours, but it's one of the fastest ways to go from zero to vCloud and to have an environment to work with vCloud Director PowerCLI. Follow the instructions in the guide provided and ensure that you get the specific versions of software noted in the guide to ensure a smooth deployment.

In addition to building a fully functioning vCloud Director implementation, AutoLab has a great collection of PowerCLI scripts that you can learn from. You can examine each of the scripts that are distributed with AutoLab to see other examples of the scripted installation and configuration of ESXi hosts.

Build your own vCloud Director implementation on your vSphere lab environment

The third option is to build and load vCloud Director in a lab environment. This method is not significantly more difficult than AutoLab, but you might run into several issues during deployment that AutoLab can handle for you. This section is an overview of the build process; however, additional instructions for each step can be located in the VMware installation documentation for each product.

For your vCloud Director installation, you will need the following software loaded in your environment:

- vCenter Server, ESXi hosts, and Microsoft Active Directory
- vShield Manager
- vCloud Director

The first step in enhancing a vCenter-managed environment and turning it into a vCloud-managed environment is to install **vShield Manager** if this isn't already in use on vCenter. vShield Manager controls network and security operations in vCenter and vCloud Director. vCloud Director requires vShield Manager, where vCenter can use it as an optional component.

Installing vShield Manager is simple. It is distributed as an OVA format virtual appliance. Perform the following steps to install vShield Manager:

1. Using vCenter Client, import the OVA and complete the custom configuration values for default passwords.

2. Once booted, log in to the web administration with the default password you defined and connect vShield Manager to your vCenter Server.

3. After connecting, make sure to navigate to the vCenter server and click on each host. Install the vShield Endpoint component and optionally the vShield App component on each host. At this point, the vShield Manager is connected and ready for vCloud Director to be installed.

The next step is to install the vCloud Director; this is straightforward, but involves a few additional steps.

1. You will need a Linux virtual machine loaded with a current operating system. CentOS is a good choice because it's free and works well with vCloud Director. You should deploy a virtual machine with two network adapters and preconfigure these network adapters with IP addresses before moving forward.

2. The next step is to install a database for vCloud Director to connect to. You can download a free version of Oracle for Linux or Microsoft SQL Express installation. Oracle for Linux can be installed on the vCloud Director VM. If you choose SQL Express, you can install this on a Windows VM (possibly the Domain Controller for very small lab environments).

3. Create a database and create the username and password needed to connect to it. You will use these credentials when configuring vCloud Director.

4. The last step before installing vCloud Director is to generate a certificate keystore that will be used for the HTTPS connection on the vCloud Director's web interface. You can generate this using the Java `keytool` command in Linux.

5. After installing a database, run the `.bin` installation file for vCloud Director in the vCloud VM. Any missing prerequisite software will be listed and you can load this with a Linux package manager.

6. Install any prerequisites and restart the installation of vCloud Director. Once it finishes the installation, it prompts you to start the initial configuration and respond back with a `yes` reply.

In the initial configuration, you will specify the addresses to use with vCloud Director and the database to connect to. The initial configuration will connect to the database and build its database and tables. Once this step completes, you can launch and log in to vCloud Director with the username and password you defined during the configuration.

vCloud Director PowerCLI doesn't allow you to easily handle the highest level of administration for the vCloud environment. Instead, you should launch it and log into the administration web interface on vCloud Director. Once logged in, follow the **Guided Tasks** section to initially configure vCloud Director against your vCenter Server and set up a new provider virtual datacenter.

Once you have completed the guided tasks in vCloud Director, you should be ready to begin the recipes of *Chapter 10, Working with vCloud Director from PowerCLI.*

Additional resources

For additional resources, refer to the following links:

- VMware vCloud Service Providers: `http://vcloudproviders.vmware.com/`
- AutoLab: `http://www.labguides.com/autolab/`
- vCloud Director Trial: `http://www.vmware.com/go/try-vcloud-director`
- vCloud Networking and Security (vShield Manager): `http://www.vmware.com/go/try-cns`
- vShield Manager Installation Guide: `http://www.vmware.com/pdf/vshield_512_quickstart.pdf`

Index

A

Active Directory
  ESXi host, joining into 7-10
add() method 183, 186
Admission Control 40
alerts
  getting, from vSphere environment 152-155
AutoLab
  about 244
  URL 244
  vCloud Director environment, deploying 244

C

**Challenge-Handshake Authentication Protocol
      (CHAP) 19**
cloning 60, 61
cluster
  advanced features, setting 39-43
  creating 37, 38
  setting, into maintenance mode with
      PowerCLI 197, 198
Comma Separated Values. *See* **CSV**
Compare-Object
  used, for auditing group
      memberships 218-220
configuration script
  creating, to set properties uniformly 25-31
ConfigureStorageDrsForPod method 105
coredump settings, for ESXi host
  configuring, from PowerCLI 185-187
CSV
  about 159
  output, sending to 159-162

custom attribute

custom attribute
  creating, number of shares per VM on
      resource pool used 140-142
custom function
  creating, for updating DRS group
      member 210-214
  used, for maintaining DRS groups
      membership 215-217
custom properties
  setting, to add context to virtual
      machines 166-168
custom storage
  configuring 182-184

D

datastore cluster
  creating 98-100
  managing 98-100
datastores
  creating, on ESXi host 18-21
Datastores view
  iSCSI 47
  NFS 47
disks
  thin to thick disks, converting with Storage
      vMotion 96, 97
disk space
  increasing, in virtual machine 73-75
Distributed Resource Scheduler (DRS)
  about 193, 194
  working with 39-43
DRS group
  creating, for virtual machines 203-205
  members, listing 206, 207

Thank you for buying
PowerCLI Cookbook

About Packt Publishing

Packt, pronounced 'packed', published its first book, *Mastering phpMyAdmin for Effective MySQL Management*, in April 2004, and subsequently continued to specialize in publishing highly focused books on specific technologies and solutions.

Our books and publications share the experiences of your fellow IT professionals in adapting and customizing today's systems, applications, and frameworks. Our solution-based books give you the knowledge and power to customize the software and technologies you're using to get the job done. Packt books are more specific and less general than the IT books you have seen in the past. Our unique business model allows us to bring you more focused information, giving you more of what you need to know, and less of what you don't.

Packt is a modern yet unique publishing company that focuses on producing quality, cutting-edge books for communities of developers, administrators, and newbies alike. For more information, please visit our website at www.PacktPub.com.

About Packt Enterprise

In 2010, Packt launched two new brands, Packt Enterprise and Packt Open Source, in order to continue its focus on specialization. This book is part of the Packt Enterprise brand, home to books published on enterprise software – software created by major vendors, including (but not limited to) IBM, Microsoft, and Oracle, often for use in other corporations. Its titles will offer information relevant to a range of users of this software, including administrators, developers, architects, and end users.

Writing for Packt

We welcome all inquiries from people who are interested in authoring. Book proposals should be sent to author@packtpub.com. If your book idea is still at an early stage and you would like to discuss it first before writing a formal book proposal, then please contact us; one of our commissioning editors will get in touch with you.

We're not just looking for published authors; if you have strong technical skills but no writing experience, our experienced editors can help you develop a writing career, or simply get some additional reward for your expertise.

Learning PowerCLI

ISBN: 978-1-78217-016-7   Paperback: 374 pages

A practical guide to get you started with automating VMware vSphere via PowerCLI

1. Automate your VMware vSphere environment including hosts, clusters, storage, and vCenter Server virtual machines and networks.

2. Create good-looking, clean reports in no time, increasing your efficiency.

3. Get to grips with PowerCLI to automate routine tasks using practical examples.

VMware ESXi Cookbook

ISBN: 978-1-78217-006-8   Paperback: 334 pages

Over 50 recipes to master VMware vSphere administration

1. Understand the concepts of virtualization by deploying vSphere web client to perform vSphere administration.

2. Learn important aspects of vSphere including administration, security, performance, and configuring vSphere Management Assistant (VMA) to run commands and scripts without the need to authenticate every attempt.

3. VMware ESXi 5.1 Cookbook is a recipe-based guide to the administration of VMware vSphere.

Please check **www.PacktPub.com** for information on our titles

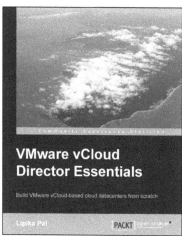